Springer Series in Language and Communication 11

Editor: W.J.M. Levelt

Springer Series in Language and Communication

Editor: W.J.M. Levelt

Language in Primates

Perspectives and Implications

Edited by
Judith de Luce and Hugh T. Wilder

Springer-Verlag New York Berlin Heidelberg Tokyo

Judith de Luce

Department of Classics
Miami University
Oxford, Ohio 45056
U.S.A.

Hugh T. Wilder

Department of Philosophy
The College of Charleston
Charleston, South Carolina 29424
U.S.A.

Series Editor:

Professor Dr. Willem J. M. Levelt

Max-Planck-Institut für Psycholinguistik
Nijmegen, The Netherlands

With 8 Figures

Library of Congress Cataloging in Publication Data
Main entry under title:
Language in primates.
    (Springer series in language and communication)
    Bibliography: p.
    Includes index.
    1. Apes—Psychology.   2. Human-animal communication.
I. de Luce, Judith.   II. Wilder, Hugh T.   III. Series.
QL737.P96L35   1983   599.8'0459   83-4502

Parts of Chapter 1 are reprinted from Annals of the New York Academy of Sciences, 1981, 364. With permission of The New York Academy of Sciences.
Chapter 8 is reprinted from *The Journal of Philosophy*, 1974, *71*, 155-168. With permission of The Journal of Philosophy, Inc. © 1974 The Journal of Philosophy, Inc.
Chapter 11 is reprinted from *The Behavioral and Brain Sciences*, 1978, *4*, 527-538. With permission of Cambridge University Press. © 1978 Cambridge University Press. Some material has been modified from the way it originally appeared.
The use of general descriptive names, trade names, trademarks, etc., in this publication, even if the former are not especially identified, is not to be taken as a sign that such names, as understood by the Trade Marks and Merchandise Marks Act, may accordingly be used freely by anyone.

Typeset by Publishers Service, Bozeman, Montana.
Printed and bound by R.R. Donnelley & Sons, Harrisonburg, Virginia.
Printed in the United States of America.

9 8 7 6 5 4 3 2 1

ISBN 0-387-90798-X Springer-Verlag New York Berlin Heidelberg Tokyo (hardcover)
ISBN 3-540-90798-X Springer-Verlag Berlin Heidelberg New York Tokyo (hardcover)
ISBN 0-387-90799-8 Springer-Verlag New York Berlin Heidelberg Tokyo (softcover)
ISBN 3-540-90799-8 Springer-Verlag Berlin Heidelberg New York Tokyo (softcover)

Preface

This anthology was originally planned in connection with a symposium "Language in Primates: Implications for Linguistics, Anthropology, Psychology, and Philosophy," at Miami University, Oxford, Ohio. Publication of the book would not have been possible without the support given to the Symposium by many individuals and groups. The Editors thank everyone involved for their kind and generous assistance. Specifically, we thank the invited speakers at the Symposium, Thomas A. Sebeok, H. Lyn Miles, Roger S. Fouts, and Thomas Simon. The chapters in this book by Miles, Fouts, and Simon are revised versions of their lectures at the Symposium. We thank Edward Simmel for his encouragement, his patience with our efforts, and his help in planning and directing the Symposium. For their financial assistance, we thank the co-sponsors of the Symposium: the Sigma Chi Foundation/William P. Huffman Scholar-in-Residence Program at Miami University, as well as the Departments of Classics, Philosophy, Psychology, and Sociology and Anthropology at Miami.

We thank Barbara Johnson, Polly J. Harris and Brenda Shaw for their secretarial and editorial help, and Shirley Gallimore for her patience, care, good humor, and hard work in typing the manuscript. Finally, we thank the contributors to this volume.

April 1983

JUDITH DE LUCE
HUGH T. WILDER

Contents

Contributors

Margaret Atherton, Department of Philosophy, University of Wisconsin-Milwaukee, Milwaukee, WI 53201 U.S.A.

Martin Benjamin, Department of Philosophy, Michigan State University, East Lansing, MI 48824 U.S.A.

Roger S. Fouts, Department of Psychology, Central Washington University, Ellensburg, WA 98926 U.S.A.

Donald R. Griffin, Rockefeller University, New York, NY 10021 U.S.A.

H. Lyn Miles, Department of Sociology and Anthropology, University of Tennessee at Chattanooga, Chattanooga, TN 37401 U.S.A.

Robert Schwartz, Department of Philosophy, University of Wisconsin-Milwaukee, Milwaukee, WI 53201 U.S.A.

Edward C. Simmel, Department of Psychology, Miami University, Oxford, OH 45056 U.S.A.

Thomas W. Simon, Department of Philosophy, University of Florida, Gainesville, FL 32611 U.S.A.

Sarah Stebbins, Department of Philosophy, Douglass College, Rutgers University, New Brunswick, NJ 08903 U.S.A.

William C. Stokoe, Linguistics Research Laboratory, Gallaudet College, Washington, DC 20002 U.S.A.

H. S. Terrace, Department of Psychology, Columbia University, New York, NY 10027 U.S.A.

Introduction

Judith de Luce and Hugh T. Wilder

The ability of apes to communicate with each other and with humans has held great popular fascination for many years. This fascination is motivated in part by curiosity about animal mentality and the comparability of human and nonhuman cognitive and intellectual capacities. Language and communication are often seen as the key to mentality, both human and nonhuman: since human language reveals so much about human thought, perhaps we can understand animal mentality if we can understand how we communicate with animals and how animals communicate with each other. More specifically, perhaps by training nonhuman animals in the use of human languages we will be better able to understand cognitive and intellectual capacities in both humans and nonhumans. This project would shed light not only on the nature of language and cognitive and intellectual capacities, but also on such issues as the uniqueness of human language and thought.

Progress made in the last few generations in linguistics, communications theory, semiotics, ethology, and developmental and comparative psychology has allowed for increasingly rigorous and fruitful study of the connections between communication, language, and cognitive capacities in humans and animals. The attempts to train chimpanzees and at least one gorilla and an orangutan in the use of a human language have been at the center of this study of human and nonhuman communication, language, and cognition. Early uncontrolled attempts to teach chimpanzees spoken English gave way to controlled experiments designed to teach apes sign language and other communication systems. Claims about the success of these experiments as well as methodological and substantive criticism of the experiments reached something of a peak in 1980. The time now seems right for an evaluation of the significance of the experiments for a number of different disciplines, as well as for speculation about fruitful directions in which this research may be pursued in the future. The aim of this anthology is to provide such an evaluation and to offer speculation about future research on language in nonhuman primates.

The first three contributions in the current volume—those by Terrace, Miles, and Fouts—are reports of recent developments in the primate language experiments, written by the principal investigators themselves. Terrace reports on his widely publicized experiment with Nim Chimpsky and levels severe criticisms against his own and others' attempts to teach apes a sign language. Miles and Fouts reply to Terrace's criticisms and describe their own experiments. The importance of the primate language studies in understanding the evolutionary development of linguistic behavior unites the essays by Fouts and the one following, by Simmel, who discusses language and communication in primates from the point of view of behavior genetics.

Simmel's interest in the wider concept of communication—as opposed to communication in human language—leads into the more philosophical essays by Stebbins, Simon, and Benjamin. Stebbins discusses the nature of communication in the primate language studies, and pinpoints some of the difficulties in earlier anthropocentric accounts of communication. The abilities of apes to communicate lead Simon and Benjamin to discussions of the ethical status of nonhuman primates and of connections between language use and moral agency.

Motivating these discussions of ethical issues are assumptions about human uniqueness and the contribution of the human use of language to this uniqueness. The putative success of the primate language experiments has often been interpreted as casting doubt on the species specificity of language in humans, and this in turn is interpreted as casting doubt on the uniqueness of human cognition and intelligence. Atherton and Schwartz discuss these questions of the species specificity of language in humans and relate them to the issue of nativism in language acquisition.

Finally, the chapters by Stokoe and Griffin strike an optimistic note about prospects for future research built on the primate language studies. Stokoe argues that the studies may be seen as contributing to the understanding of symbolic communication in animals, including humans, and that therefore the studies contribute to the development of the science of semiotics. Griffin stresses the place of the studies in the developing field of ethology, and argues that the primate language studies may provide a new "window" onto the world of thought in animals.

The questions raised by the research on language in nonhuman primates range from rather grand philosophical issues, such as the connection between language and human nature, to more precise empirical issues, such as the ability of chimpanzees to master elements of a sign language. In the remainder of this essay we introduce some of the issues raised in the primate language studies through a critical review of the contributions to this volume.

Linguistic Media and Training Methods

The attempts of the Kelloggs (Kellogg, 1968) and the Hayeses (Hayes, 1951) to teach chimpanzees spoken English represent the first scientific efforts in the United States to train apes in the use of a natural human language. These early attempts were unsuccessful, some believe, due to the differences between human and chimpanzee vocal organs. It was conjectured that the chimpanzee's failure to acquire spoken English was due to anatomical rather than cognitive factors. When Gardner and Gardner (1969, 1971, 1974, 1975, 1978), in Project Washoe, continued the effort to teach a chimpanzee a human language, the language selected was American Sign Language (ASL), the visual language of many deaf people in the United States. ASL was seen as desirable because it is an established language with fluent users who could participate as investigators and assistants in the experiments, and because chimpanzees are able to form the signs of ASL.

While the linguistic medium was new, in other crucial respects the experiments of the Kelloggs and Hayeses serve as exemplars in one research program that is being pursued in primate language experiments. This research program, typified in the work of Gardner and Gardner, is being carried on, with variations, by Fouts and Miles as well as others (see especially Patterson, 1978a, 1978b; Patterson & Linden, 1981). It is characterized by several features: (1) it proceeds on the assumption that the most appropriate linguistic medium for the experiments is a language in everyday use among humans; (2) it assumes that the appropriate learning environment for the primate subjects is comparable to the natural home environment in which children acquire language; (3) the primary objective is the comparative psychological study of the cognitive development of human and nonhuman primates; and (4) emphasis is on the study of cognitive development generally, rather than on the final acquisition in the subjects of a communication system that meets the necessary and sufficient conditions of a "natural human language," whatever those may turn out to be.

The research program represented in this collection has made significant contributions to comparative psychology. Investigators claim dramatic success in teaching nonhuman primates the conversational use of ASL. Perhaps more importantly, the research has provided a new approach to the study of comparative psychology. As we will see, from the vantage point of this wider interest in comparative developmental and cognitive psychology, "failure" at any point in training animals in the use of human language is as instructive as "success."[1]

The research program described by Miles and Fouts in Chapters 2 and 3, respectively, is to be distinguished from work (e.g., that of Premack, and Savage-Rumbaugh and Rumbaugh) that attempts to train chimpanzees in the use of various artificial languages. It is also to be distinguished from work that emphasizes controlled experimental environments and regimes of operant conditioning at the expense of exposure to language in homelike environments. Investigators using artificial systems, such as Premack's plastic chips and Savage-Rumbaugh and Rumbaugh's "lexigrams," criticize the research using ASL in homelike environments as being uncontrolled and as yielding merely anecdotal results (as reported by Marx, 1980). It should be noted that while Terrace levels similar criticisms against several ASL experiments, his own Project Nim, as reported in Chapter 1, did use ASL in what he describes as a "playful and spontaneous"

---

[1] See pp. 9 and 15.

environment (see also Terrace, 1979; Terrace, Petitto, Sanders, & Bever, 1979). The following methodological questions remain open, however. First, given that the experimental objective is to study comparative cognitive development in humans and apes, what is the most appropriate communication medium? The options include ASL or other everyday human languages, various artificial languages, or the natural communication systems used by apes in the wild. Interest in cognitive development itself does not yield a preference for any one of these three options. Second, one must also ask what training methods and rearing environments are most appropriate for the subjects. The answer to this question will depend in part on the communication medium selected: if a medium naturally occurring in either humans or apes (or resembling one naturally occurring) is being used, then one will probably want to study the acquisition of this system in environments resembling those in which they are naturally acquired. These outstanding methodological issues must be addressed in any further attempts to study human and ape cognitive development through the study of the species' acquisition of communication media.

Terrace, in Chapter 1, raises other problems concerning his own and others' attempts to teach apes sign language. These problems lead Terrace to be skeptical of all previous claims alleging that apes have acquired the ability to create sentences according to rules of grammar. This ability, Terrace believes, is essential to the use of language. Terrace has no doubt that Nim and a number of other nonhuman primates have been trained to form strings of signs of ASL, and that these strings exhibit both lexical and semantic regularities. What Terrace does doubt is the explanation of these regularities in terms of the hypothesis that the apes are creating sentences according to grammatical rules; the regularities may have another, nongrammatical explanation. The principle that warrants a preference for the nongrammatical over the grammatical explanation is known as *Lloyd-Morgan's Canon*[2]: according to it, "we must never interpret a piece of animal behavior as the outcome of a higher capacity or power of mind if it can be interpreted satisfactorily as the outcome of a lower one" (Broadhurst, 1963, p. 25). Granted that the terms "lower" and "higher" capacities or powers of mind are ill defined, it is still clear that according to this principle the nongrammatical explanation is to be preferred over the grammatical, *if* the behavior of the apes can be interpreted satisfactorily according to the nongrammatical hypothesis.

Terrace's explanation of the apes' behavior is, in general, that they are responding in noncreative and therefore nonlinguistic ways to the trainers' coaxing and cueing. Terrace supports this explanation through his statistical analysis of Nim's multisign "utterances" and in his analyses of videotapes and films of the performances of Nim, Washoe, Ally and Booee (two chimpanzees raised and

---

[2] It is not, as M. Gardner suggests, any version of Occam's Razor. Occam's Razor is a principle of ontological parsimony; Lloyd-Morgan's Canon is a principle of human intellectual hegemony. See M. Gardner (1980) for the Occam's Razor view; see Simon, this volume, pp. 99-100, for the Lloyd-Morgan's Canon view.

trained at the Oklahoma Institute for Primate Studies), and Koko (a gorilla raised by Francine Patterson at the Gorilla Foundation in California).

Clever Hans Phenomenon

The problem raised by Terrace is similar to an even more insidious problem that many think plagues and even dooms all of the primate language experiments, or at least those that attempt to train the animals in homelike social environments in the conversational use of an everyday human language. Since Rosenthal's revival in 1965 of Pfungst's *Clever Hans (The Horse of Mr. von Osten)*, this problem has been known as the *Clever Hans phenomenon*. It is "the subtle and unintentional cueing of the subject by the experimenter" (Rosenthal, 1965, p. xxii). Sebeok and Umiker-Sebeok have most carefully described how the problem might infect the primate language experiments; they believe that no experiment has yet been shown to be free of it (Sebeok, 1978; Sebeok & Umiker-Sebeok, 1979, 1980).

The possibility of trainer and/or observer prompting and cueing in these experiments may always undermine the explanation of the animals' behavior by their hypothetical possession of linguistic abilities. Rosenthal suggested that the phenomenon occurs in all experimental situations in which there is any contact between experimenter and subject, whether the subject is a human or an animal; the phenomenon is well illustrated in the case of Clever Hans himself.

Hans was a horse exhibited by a German named von Osten in Berlin at the turn of the century. Hans had been trained to answer questions put to him by observers, and his answers exhibited humanlike intelligence. Hans communicated by tapping one foot and by moving his head. He could answer correctly nearly all the questions put to him in German. Hans seemed to be able to count, solve problems in arithmetic, and read German; he seemed to have an excellent memory (for example, he could correlate days and dates in a given calendar year); he could tell time; he could recognize faces; he had as keen an eye and ear as a human; he possessed perfect tone consciousness, and so on and on.

Observers were amazed by Hans' abilities, and many concluded that Hans had proved that animals possessed and could develop the same kind of intelligence as humans. Others were skeptical, and scientists as well as nonscientific observers at Hans' exhibitions attempted to explain his abilities by searching for some form of communication between Hans and von Osten. Most of these skeptics assumed that von Osten was deceiving the observers and was deliberately controlling Hans' tapping responses in some subtle way. However, no one could discover any deliberate communication or means of control von Osten might have over Hans.

In 1904, however, Pfungst and Strumpf began to look for unconscious and nondeliberate communication between Hans and his questioners. Pfungst noticed that Hans could answer questions in the absence of von Osten, which rendered improbable any explanation in terms of communication or control by von Osten. However, Pfungst also noticed that Hans' abilities diminished when he could not see his questioner, and when the questioner himself did not know the answer to the questions. Pfungst hypothesized that Hans' abilities depended not on his own apparently human intelligence, but on as yet undiscovered visual cues given by the questioner.

Of course, not all of Hans' questioners could be witting participants in a fraud, deliberately passing Hans information or controlling his behavior. Pfungst therefore looked for visual cues that might be passed unwittingly from questioner to horse. Pfungst finally discovered these cues in the unintentional and largely involuntary motions of the questioner in asking his question and watching Hans answer. Relevant motions included changes in the posture and inclination of the head of the questioner, raising of eyebrows, and dilation of nostrils. Certain subtle motions were made indicating expectancy when the questioner asked his question, and these motions cued Hans to begin tapping. Other motions were made indicating relief or satisfaction when the questioner thought Hans had tapped enough times to answer his question, and these motions cued Hans to stop tapping. What Hans had learned to do was to read these nearly subliminal cues in his questioners' behavior; he had not learned arithmetic, German, how to tell time, etc. When observers had perceived humanlike intelligence in Hans' behavior, what they were actually seeing was the human intelligence of Hans' questioners projected onto the animal.

Rosenthal (1965, p. xiii) observed that "Pfungst's findings solved not only the riddle of Clever Hans but in principle the problem of other 'clever' animals." Sebeok pointed out that, in particular, Pfungst provided the solution to the problem of explaining the behavior of the "clever" apes who have learned the signs of ASL. Sebeok's point is that these apes have in fact also learned something else—which is not the ability to create sentences according to grammatical rules of ASL: they have learned how to read the signs of their masters.

Sebeok's argument does not depend on the primate language experimenters knowingly and deliberately cueing their subjects. Pfungst did not claim that Hans' questioners deliberately cued Hans; in fact, his argument is that the cues of Hans' questioners were *un*intentional, and this is precisely what makes them so interesting as a medium of communication. This is also what made them hard to detect, however. Although Sebeok (and Terrace as well) claims to have detected cues in the behavior of the investigators in the ape language experiments, his claim is not that there is an obvious and known system of cues at work. His claim is, rather, that the cues *must* be there, whether or not the investigators or anyone else recognizes the cues and can identify them. The reason the cues must be there is because the experiments—especially those in the Kellogg-Hayes tradition—depend on close social interaction between investigator, assistants, and subject. As Rosenthal pointed out, investigator expectation is an ineradicable factor affecting the result of any experiment involving interaction between investigator and subject. This is so, Rosenthal argued, whether the subject is a rat in a maze or a child in a classroom. Investigators signify their expectations in complex systems of unintentional and involuntary cueing behavior and their behavior affects their interaction with the subjects, whether the subjects are pigeons they are placing in Skinner boxes or chimpanzees with whom they are conversing in ASL. The behavior of the subject is affected by the expectations of the experimenter, as these expectations are communicated in the experimenter's cues.

Clever Hans and Primate Language Experiments

How the Clever Hans phenomenon might affect the primate language experiments and the validity of the results of the experiments are vexed and complicated questions. These questions are addressed by Fouts, Miles, Simon, Stebbins, Stokoe, and Griffin in this volume. It is important to recognize that to claim that the Clever Hans phenomenon of trainer cueing has not been eliminated from the primate language experiments is not to accuse the investigators of deliberate fraud or deception. The possibility of fraud was a primary motivation in Pfungst's analysis of Clever Hans, but his explanation of Hans' behavior was precisely what exonerated von Osten in the case.

If there is any deception in the Clever Hans case or in the primate language experiments, say the skeptics, it is a type of self-deception on the part of the investigators. Looking for humanlike intelligence in the apes, the investigators have found it; what they have deceived themselves into believing, it could be claimed, is that the apes in fact possess this intelligence. The apes' intelligence may be simply a projection of the investigators' own.

This skeptical argument depends on two suppressed premises: Lloyd-Morgan's Canon, and a strong substantive claim about the nature of human language. It is important to recognize how these premises function in the argument. First, remember that Lloyd-Morgan's Canon authorizes explanation of the apes' behavior in terms of responses to unintentional cues rather than the creative use of language only if the creative use of language is a "higher" capacity than responding to the sorts of unintentional cues involved in the Clever Hans phenomenon.

It is at this point that the assumption about the nature of language enters the argument. Terrace acknowledges the premise explicitly: Chomsky and Miller "have convincingly reminded us," he writes, "of the futility of trying to explain a child's ability to create and understand sentences without a knowledge of rules that can generate an indeterminately large number of sentences from a finite vocabulary of rules" (this volume, p. 19). The assumption is that this creative rule-governed use of language is a higher capacity than responding to Clever Hans-type paralinguistic cues. This assumption itself only makes sense if we in fact have two distinct capacities here, one of which could be "higher" than another. This distinction is questionable, however, on the basis of how humans actually use language in communication. The fact is that humans could not use language at all were it not for the presence of Clever Hans-type cues in conversational situations. This point applies to the use of language in adults as well as the use of language in children as they are acquiring it.

A related point has often been made in connection with Chomsky's emphasis on the role of formal rules of syntax in the generation of sentences: while crucial in explaining formal properties of grammar, these rules are insufficient (and are not intended to be sufficient) in explaining pragmatic features of communication in language. Communication in language, and development of the ability to communicate in language, depend on the presence of nongrammatical and nonlinguistic cues in conversational situations.

Two relevant implications follow from these observations. First, if the objective of the primate language experiments is to compare the cognitive development of apes and humans, the presence of the Clever Hans phenomenon in the ape experiments does not itself invalidate the experiment, since the Clever Hans phenomenon is present in human interaction as well (see Stokoe, this volume, pp. 151-152). Second, if Lloyd-Morgan's Canon in combination with the Chomskyan necessary condition for language authorizes the deflationary interpretation of the apes' signing behavior, then it authorizes a deflationary interpretation of human linguistic behavior as well: one of the primary skills involved in being able to communicate in language—viz., to be able to respond appropriately to Clever Hans-type cues—is so simple even an ape can do it. This argument need not be taken as a reductio ad absurdum of the skeptic's position; this deflationary account of human intelligence and linguistic abilities seems to be just what some behaviorist psychologists take to be correct. This deflationary account might also tally with certain biological accounts of communication within and across a variety of species, as these accounts stress continuities in the communication systems of various species. (Perhaps surprisingly, see Fouts' account along this line in Chapter 3.)

A final point about the Clever Hans phenomenon is that discovery of this phenomenon in cases of intelligent animal behavior does not by any means settle the question of the similarity, in kind or degree, of animal intelligence to human intelligence. This point was made carefully by Pfungst himself in his report on Clever Hans. Pfungst cautioned against generalizing about "animal intelligence" on the basis of his explanation of Hans' behavior and even on similar explanations of all "clever" animals (Rosenthal, 1965, pp. 241-242). All of these animals— including the chimpanzees, orangutan, and gorilla in the primate language studies —are domesticated animals, socialized to live with human beings. Pfungst pointed out that it is possible that these "animals have suffered in the development of their mental life, as a result of the process of domestication" and experimentation, and quoted with approval Buffon: *"Un animal domestique est un esclave dont on s'amuse, dont on se sert, dont on abuse, qu'on altère, qu'on dépaise et que l'on dénature"* (Rosenthal, 1965, pp. 241-242).[3] While most investigators have assumed that the primate language experiments develop and enhance the intelligence and linguistic abilities of the apes, just the opposite may be the case.

---

[3] A domestic animal is a slave with which one amuses oneself, which one makes use of, which one abuses, corrupts, removes from its natural element and perverts.

Cognitive and Communicative Abilities in an Orangutan

Pfungst was warning investigators about the dangers of drawing generalizations about animal intelligence on the basis of limited, and perhaps contaminated, evidence. Miles addresses both points in her contribution to this volume. The Director of Project Chantek, which is attempting to train an orangutan in the use of ASL, Miles writes that "my goal in raising Chantek was to provide a comfortable environment, but not to attempt to rear him totally like a human child. I wanted to retain natural orangutan behaviors and provide at least a partially arboreal environment for him to develop as normally as possible, within the constraints of our research program" (this volume, p. 47). Miles presumably adopted this goal because of the methodological superiority of conducting ape language studies in ways that enhance rather than detract from the naturally occurring intelligent and communicative behavior of the animals. This methodological point is further explored in the chapters by Fouts, Stebbins, and Simon.

Miles has adopted what she calls an "analytical and developmental approach" in her research program. Rather than taking as her objective the development in Chantek of behavior that meets the latest set of necessary and sufficient conditions for true language use, Miles' concern is to describe and analyze the development of Chantek's cognitive and communicative abilities. She is interested in various patterns of social and cognitive development that are associated with different degrees of success in acquiring language, and stresses similarities between her approach with Chantek and recent studies of language acquisition in children. She cites with approval those integrative studies that assume relationships among social, cognitive, communicative, and linguistic development.

Miles believes that her work with Chantek supports "the evaluation of the linguistic abilities of apes as comparable to the performance of human children in the earliest stage of language acquisition." That apes can achieve such a level is disputed, as we have seen, by Terrace. Miles compares Project Chantek with Terrace's Project Nim, and concludes that "the failure of Terrace and his colleagues to teach Nim communicative and linguistic skills should be viewed as the failure of one training method, not of the capacity of apes" (this volume, p. 59). This conclusion echoes the cautionary note of Pfungst about the dangers of generalizing about animal intelligence on the basis of one "failure" of intelligence. Miles' conclusion also illustrates the spirit that still animates the "animal intelligence" debate.

Continuities Between Communicative Abilities in Human and Nonhuman Animals

Fouts also criticizes Terrace's Project Nim and offers an account of his own theoretical orientation in the primate language studies. His orientation is similar —in its stress on continuity between social, cognitive, communicative, and linguistic abilities in human and nonhuman animals—to the orientation of Miles just described. Fouts adds an emphasis on evolutionary continuity in communicative behavior.

Fouts begins Chapter 3 by detecting his own version of the Clever Hans phenomenon in the primate language studies: the varying results of the studies tell us more, he claims, about the expectations of the investigators—as embodied in their theoretical assumptions about the nature of animal behavior—than about the capacities of the animals. On this basis, Fouts criticizes the studies of Terrace and others. Fouts claims that Terrace's methodology in Project Nim ignored crucial social and conversational conditions of linguistic ability; Terrace's theory of linguistic behavior, deriving (ironically) from both Skinner and Chomsky, also ignores these social and conversational conditions; and, not surprisingly, Fouts concludes, Nim himself failed to acquire these social and conversational underpinnings of linguistic ability. Here is a case, Fouts claims, of "failure" being as instructive as success: Nim's failure instructs "in the same [way] that a deprivation experiment tells what is important for the natural development of a behavior" (this volume, p. 64). That linguistic or at least communicative behavior does in fact naturally develop in primates is obviously an assumption of Fouts' theory of animal behavior. True to his own version of the Clever Hans phenomenon, when Fouts looks for linguistic or communicative behavior in animals trained according to his integrative methods, he finds it.

Fouts includes a description of his present research on the question of whether one chimpanzee might be able to teach another some linguistic skills in ASL. Fouts sees this as a promising line of research: a mother chimpanzee who has already acquired some ASL might be able to teach her infant some of the skills. Fouts thinks that is possible because such research builds on the fact that linguistic activity is social and develops normally in the context of socially significant (e.g., parent-child) relationships. In general, he argues, if it is to occur at all,

> language acquisition in nonhuman primates must begin in infancy in the context of strong social bonds and utterances must be made in contexts meaningful to a young organism. Furthermore, the medium of communication should be compatible with the biology of the primate and, specifically, should take advantage of the predisposition of apes to communicate with gestures. (this volume, p. 73)

The same points are explored, as has already been noted, by Miles and, as we shall see, by Stebbins and Simon in their chapters in this volume.

The concepts in terms of which Fouts describes communicative behavior—e.g., "sequential" and "simultaneous processing"—apply to both human and nonhuman behavior. Fouts argues, in effect, that such conceptual and biological continuities between communication in human and nonhuman primates have an evolutionary basis. Simmel, in Chapter 4, pursues this line of investigation, probing the evolutionary basis of communicative behavior in apes and humans. In general, Simmel argues that a behavior-genetic approach may help illuminate the abilities of nonhuman primates to acquire a sign language.

Simmel asks the inevitable question: if apes are capable of language, why haven't they developed that means of communication on their own, in their natural environment? In Chapter 4, Simmel explores several answers to this question. He regards the ape language studies as an opportunity to consider the role of complex behaviors in evolution and to further the study of the communication process itself.

After discussing natural selection as it operates on phenotypes, Simmel acknowledges a number of difficulties encountered when trying to study language-using apes in terms of evolution (e.g., the paucity of apes available for such study, and the length of time needed for an ape to attain reproductive maturity). Nevertheless, Simmel asserts that it is possible to look at some important aspects of language-related phenotypes. While primary interest in the language studies has tended to focus on cognition, he suggests that it might be important to broaden the investigation to include other variables, such as emotions and social factors, which probably bear on the evolutionary significance of language. Moreover, Simmel perceives a need to develop phenotypic measures that go beyond acquisition, such as the readiness with which a subject matter is acquired or the degree of spontaneity in signing.

At the beginning of Chapter 4, Simmel asks in passing whether the human species survived because, having failed with simpler forms of communication, it had developed a more complex form, language, which contributed to the species' reproductive success. Simmel concludes by questioning whether we would discover any effects on reproductive success were there to develop a sufficiently large population of language-using apes to allow such a study.

Both Fouts and Simmel concern themselves with the evolutionary basis of continuities between communicative behavior in apes and humans. One way to understand the debate about the primate language studies is to see the skeptics as challenging the view that the type of communication of which apes are capable is anything like the type of communication *in language* of which humans are capable. As Stebbins points out in Chapter 5, the traditional (perhaps Cartesian) assumption has been that communication in language is a special case of communication, although discontinuous (both logically and biologically) with the other cases. We have raised and shall return to the question of "language" in primates, but shall stay with Stebbins for a moment on the more basic question of "communication" in primates.

Stebbins notes that if the criticisms of Terrace and Sebeok are sound, the apes may not even be communicating in ASL. They may simply be "responding to nonverbal signals from their trainers, or they may be exhibiting conditioned responses to the presence of an object or in the expectation of a desired reward" (this volume, p. 85). Stebbins does not give a direct answer to the question of whether the apes are communicating; rather, she illustrates the mistakes that can occur when researchers make claims about "the nature of communication" based solely on parochial assumptions applicable to a few local cases of (usually human linguistic) communication. Her point is that while there may be continuities between types of communication in various species, still "different features may characterize the processes and products" of cases of communication in different species. Any definite answer to the question of whether the apes are communicating in the ape language experiments is likely to tell us more about our anthropomorphic understanding of communication than about the apes' behavior.

Methodological and Ethical Issues in Primate Language Studies

Simon, in Chapter 6, returns to the issue of whether the apes have been shown to use language in the experiments. Simon considers several methodological objections (including one version of the Clever Hans phenomenon) to the experiments, each of which he concludes fails to make its case for the skeptical claim that the apes have not been using language. Simon does not argue that the apes *do* use language; his point is that the methodological objections do not invalidate this conclusion.

Simon also discusses the ethics of primate language research. Three general points are made. First, an irony of the primate language studies is that the more successful they are in demonstrating the humanlike intelligence and linguistic capacities of apes, the more questionable the experiments are from an ethical point of view. Second, Simon argues that the procurement processes and treatment of the apes in the laboratory setting inflict harm on the apes. Finally, if alternative, less harmful approaches are available to the present primate language studies, then they are preferable from an ethical point of view. Obviously, whether alternative approaches are in fact available depends on both the objectives of the experiments as well as the practical problems involved in the alternatives. Simon argues that if the experimental objective is to study processes involved in understanding natural languages, then computers may be better experimental subjects than apes. If understanding language development in humans is the experimental objective, then children should be studied directly: problems of extrapolation will be eliminated, and the ethical questions will be impossible to avoid. If the comparative study of animal and human intelligence is the goal, as we have seen it is for some of the investigators, then "field" studies of both animals and humans is to be preferred, on both ethical and methodological grounds.

In claiming that the more successful the primate studies are in demonstrating linguistic abilities in apes, the more questionable the studies are from an ethical point of view, Simon is assuming a close connection between possession of language and possession of moral rights and values. In Chapter 7 Benjamin considers the problem of the connection between possession of language and possession of moral rights, as well as the related problems of the moral status of animals and human obligations to animals.

Benjamin considers three historical positions on the obligations humans might have to animals. The first position, represented by Aquinas and Kant, is that human beings have only indirect obligations to other animals. Kant maintained that self-consciousness is necessary in order to be the object of a direct obligation, and in Kant's view only humans are self-conscious. The second position, taken by Descartes, maintains that humans have no obligations to nonhuman animals. Descartes, judging animals to be living "machines," regarded the capacity to use language as the feature that distinguishes living human bodies that are "ensouled" from those animal bodies that are not. Finally, Bentham represents the third position, which claims that humans have direct obligations to animals, who share with humans the capacity to experience pleasure and pain.

Benjamin goes on to propose a fourth position, which he calls "nonutilitarian direct obligation." Benjamin believes that whether or not animals can use language is not an ethically significant issue. To be taken seriously from the moral point of view, animals merely have to be sentient. It is important, however, that humans qua persons have language because language use establishes their special worth as well as their special obligations. Benjamin concludes with a discussion of the connection between language and personhood, asking what nonhuman primates would have to show to acquire significance in terms of ethics, i.e., personhood. He argues that animals must be thought to communicate with themselves if they are to be regarded as persons; as persons, in his technical sense, they would then accrue both rights and obligations.

Linguistic Nativism and the Species Specificity of Language

Benjamin's chapter illustrates well the point that claims about the moral uniqueness of humans are often founded on further claims about the uniqueness of human linguistic capacity. The species specificity of language has been debated since antiquity and the primate language studies clearly have implications for this debate. Atherton and Schwartz discuss these implications in their two contributions to this volume. They carefully distinguish between "linguistic nativism" and claims about the "species specificity of language" in humans. Linguistic nativism is the view that human linguistic competence is the product of innate factors. The view that language is specific to the human species asserts that linguistic competence is unique to humans. While linguistic nativism and claims about the species specificity of language are distinct, they are often confused in the literature, and the first has been thought to imply the second. Atherton and Schwartz point out that this implication does not hold.

Their main purpose in Chapter 8 is to explore the underlying assumptions and the implications of the nativist position, in order to assess the extent to which animal studies may provide an empirical basis for countering that position. Atherton and Schwartz argue that nativism, in order to be intelligible as a theory of language development, must be part of a learning theory in which mental structures are cited to explain the human ability to learn language. The authors distinguish between claims that the ability to learn is due to innate factors and claims that that which is learned is itself innate. They suggest that animal studies may illuminate the nativist claim that mental structures are task specific. Caution must be exercised in interpreting the significance of the language experiments, however, particularly because it is difficult to separate the evaluation of the capacity to learn language from the evaluation of intelligence in general.

Atherton and Schwartz conclude that the prospects for linguistic nativism are unclear. They suggest that if the factors for language competence actually differ from other factors bearing on human abilities to make abstractions, discriminations, etc., then it may not be reasonable to argue that natural language is species specific and accounts for human nature. More importantly, by claiming that the factors for natural language are task specific and separable from other aspects of cognition, the nativists would remove the major philosophical interest from the claim that natural language is innately species specific.

While the recent controversy over primate language studies has tended to dwell on what several apes appear to be able to do, Atherton and Schwartz believe that considerably more attention needs to be paid to the significance of determining the linguistic capacities of nonhuman primates. The language studies have been viewed as tests of the genetic basis of language acquisition, linked in traditional thought to claims for the uniqueness of human mentality. If apes can learn language, then the claims for a genetic basis for a unique human mentality are undermined.

Atherton and Schwartz doubt, however, that the results of the language studies can have much bearing on the question of human language acquisition. Reminding the reader that uniqueness and innateness are not the same, Atherton and Schwartz suggest in Chapter 9 that ape linguistic success may be irrelevant to an innateness hypothesis since the reasons for ape failure to learn a language may have little to do with genetic cognitive endowment. Moreover, ape failure would not necessarily reflect on human success (but see Fouts, Chapter 3, for a rebuttal of this claim).

Atherton and Schwartz are skeptical of the heavy emphasis put on syntactical skills in the primate language studies. They are not convinced that ape success or failure at syntax reveals much about ape mentality. Success at syntax, for example, does not reveal the presence of creativity or interesting thoughts. Moreover, it is not necessary to prove that animals can acquire a natural language syntax in order to demonstrate possible symbolic activity, which would in turn implicate mind.

Since one judges mentality by the content of the messages conveyed as well as by linguistic skill, current tests of language-using apes (e.g., tasks to label objects) are unlikely to elicit information that reveals interesting thoughts. In addition, to reveal thought, we need evidence that the individual understands what he or she is talking about. In Atherton and Schwartz's estimation, the ape studies have been inconclusive because the performance of nonhuman primates in these experiments can neither prove that ape minds are not mechanistic nor prove that human minds are unique.

One of the issues in the ape language studies, then, is continuity versus discontinuity in ape-human mentality and linguistic ability. Atherton and Schwartz report inconclusive results in this debate. Stokoe, in Chapter 10, argues that the science of semiotics may offer a systematic and comprehensive framework for understanding the ape language studies and communication in general. Semiotics, applied to the ape language studies, may reveal important underlying continuities between communication in apes and humans.

The status of ASL is often misunderstood by both investigators and critics of the primate language studies. ASL is not "signed English"; it is a language with its own syntax and semantics. One of Stokoe's main points is that once ASL is understood to be the language it is, then continuities will be recognized between it and other languages, and between it and other more basic systems of signification and communication. Continuities may also be studied in the comparative development of human infants acquiring their first language and young apes acquiring ASL. In this way, Stokoe's understanding of ASL and of the connections between sign languages and related but distinct systems of signification and communication lead him to see great potential value in the primate language studies for the science of comparative psychology.

Stokoe also discusses the Clever Hans phenomenon and its role in the primate language studies. Given an interest in communication, Stokoe points out that the presence of the Clever Hans phenomenon in the primate language studies enhances rather than diminishes the interest of the studies. The Clever Hans phenomenon is of interest itself because it is a dramatic and unexpected case of communication between humans and a horse. Therefore, the primate language studies may provide a richer forum than might be expected for the investigation of cross-species communication.

Animal Communication as a "Window"

In order to understand human uniqueness more fully, it is appropriate to study the nature of animal thinking. In spite of the fact that animal behavior suggests that other animals do make choices, anticipate future events, and understand simple relationships and concepts, behavioral scientists have tended to ignore the possibility that nonhuman animals know what they are doing. Griffin, in Chapter 11, proposes a cognitive ethology whose goal is to study the likelihood that nonhuman animals have mental experiences, the nature of those experiences, and their effect on the animals' behavior. In his view, this can be done without violating the materialist perspective of behavioral scientists, psychologists, and ethologists, for to grant animals awareness does not require granting them an immortal soul as well.

Griffin believes that animal communication experiments can be especially illuminating by analogy with human language. Human language provides a unique "window" onto human thought and is generally assumed to be an essential element in thinking. It has been assumed, in addition, that only humans have the capacity for language and thus alone are thinking animals. Animal communication, on the other hand, is regarded as distinctly different from human language, in part because animals have traditionally been denied the conscious intent to communicate. Much of human thought, however, is expressed non-verbally, and ethologists have observed considerable variety in nonverbal communication in many species. If language reveals human thought, then nonverbal communication must be recognized as revelatory of animal thought.

By the same token, whatever linguistic communication animals are shown to be capable of in the primate language studies may reveal features of animal awareness and cognition. Furthermore, continuities between human thought and animal thought may be revealed through studies of human and animal communication. As Griffin concludes, "Animal awareness, if it occurs, is also important for our definition and understanding of the human condition" (this volume, p. 183). The primate language studies offer a means of answering the questions "how do animals understand us, and how can we understand animals?" (Stokoe, this volume, p. 152). To echo Griffin, perhaps the primate language studies can help answer another question: How can we understand ourselves?

References

Broadhurst, P. L. *The science of animal behavior*. Baltimore: Penguin Books, 1963.

Gardner, M. Monkey business. *New York Review of Books*, March 20, 1980, *27*, 3-6.

Gardner, R. A., & Gardner, B. T. Teaching sign language to a chimpanzee. *Science*, 1969, *165*, 644-672.

Gardner, R. A., & Gardner, B. T. Two-way communication with an infant chimpanzee. In A. M. Schrier & F. Stollnitz (Eds.), *Behavior of non-human primates* (Vol. 4). New York: Academic Press, 1971.

Gardner, B. T., & Gardner, R. A. Comparing the early utterances of child and chimpanzee. In A. Pick (Ed.), *Minnesota symposia in child psychology*. Minneapolis: University of Minnesota Press, 1974.

Gardner, R. A., & Gardner, B. T. Early signs of language in child and chimpanzee. *Science*, 1975, *187*, 752-753.

Gardner, R. A., & Gardner, B. T. Comparative psychology and language acquisition. *Annals of the New York Academy of Science*, 1978, *309*, 37.

Hayes, C. *The ape in our house*. New York: Harper & Row, 1951.

Kellogg, W. N. Communication and language in the home-raised chimpanzee. *Science*, 1968, *182*, 423-427.

Marx, J. L. Ape-language controversy flares up. *Science*, 1980, *207*, 1330-1333.

Patterson, F. G. The gestures of a gorilla: Language acquisition in another pongid. *Brain and Language*, 1978, *12*, 72-97. (a)

Patterson, F. G. Linguistic capabilities of a lowland gorilla. In F. C. C. Peng (Ed.), *Sign language acquisition in man and ape: New dimensions in comparative psycholinguistics*. Boulder, Colo.: Westview Press, 1978. (b)

Patterson, F. G., & Linden, E. *The education of Koko*. New York: Holt, Rinehart & Winston, 1981.

Rosenthal, R. (Ed.). *Clever Hans (The horse of Mr. von Osten), by Oskar Pfungst.* New York: Holt, Rinehart & Winston, 1965.

Sebeok, T. A. Looking in the destination for what should have been sought in the source. In T. A. Sebeok, *The sign and its masters.* Austin: University of Texas Press, 1978, pp. 85-106.

Sebeok, T. A., & Umiker-Sebeok, J. Performing animals: Secrets of the trade. *Psychology Today*, November 1979, 78-91.

Sebeok, T. A., & Umiker-Sebeok, J. Questioning apes. In T. A. Sebeok & J. Umiker-Sebeok (Eds.), *Speaking of apes.* New York: Plenum Press, 1980.

Terrace, H. S. *Nim.* New York: Knopf, 1979.

Terrace, H. S., Petitto, L. A., Sanders, R. J., & Bever, T. G. Can an ape create a sentence? *Science,* 1979, *206*, 891-902.

CHAPTER 1

Apes Who "Talk": Language or Projection of Language by Their Teachers?

H. S. Terrace

In 1917, Franz Kafka published a story ("A report to an Academy") about a chimpanzee who acquired the gift of human language. During the years that have elapsed since the publication of Kafka's tale, much has been written, on the one hand, about man's presumably unique capacity to use language and, on the other hand, about real apes whose human teachers claim that they have mastered certain features of human language. After attempting to teach my own chimpanzee to use a human language, I questioned these claims on the very grounds that Kafka perceived to be the basis of his fictional chimpanzee's ability to talk:

> ... there was no attraction for me in imitating human beings. I imitated them because I needed a way out, and for no other reason ... And so I learned things, gentlemen. Ah, one learns when one needs a way out; one learns at all costs. (Kafka, 1917/1952, pp. 178–179)

In this chapter I review the grounds for my skepticism about an ape's ability to learn a human language.

The question, "What is language?", has yet to be answered by linguists, psychologists, psycholinguists, philosophers and other students of human language in a way that captures its many complexities in a simple definition. They do agree, however, about one basic property of all human languages: that is the ability to create new meanings, each appropriate to a particular context, through the application of grammatical rules. Chomsky (1957) and Miller (1964), among others, have convincingly reminded us of the futility of trying to explain a child's ability to create and understand sentences without a knowledge of rules that can generate an indeterminately large number of sentences from a finite vocabulary of words.

The dramatic reports of Gardner and Gardner (1969), Premack (1970), and

Rumbaugh (Rumbaugh, Gill, & Glaserfeld, 1973) that a chimpanzee could learn substantial vocabularies of words of visual languages and could also produce utterances containing two or more words, raise an obvious fundamental question: Are a chimpanzee's multiword utterances grammatical? In the case of the research by Gardner and Gardner, one wants to know whether Washoe's signing *more drink* in order to obtain another cup of juice, or *water bird*, upon seeing a swan was creative juxtaposition of signs. Likewise, one wants to know whether Sarah, Premack's main subject, was using a grammatical rule in arranging her plastic chips in the sequence, *Mary give Sarah apple*, and whether Lana, the subject of a related study conducted by Rumbaugh et al., exhibited knowledge of a grammatical rule in producing the sequence, *Please machine give apple*.

In answering these questions, it is important to remember that a mere sequence of words does not qualify as a sentence. A string of words learned by rote presupposes no knowledge of the meaning of each element and certainly no knowledge of the relationships that exist among the elements. Sarah, for example, showed little, if any, evidence of understanding the meanings of *Mary, give,* and *Sarah* in the sequence, *Mary give Sarah apple*. Likewise, it is doubtful that, in producing the sequence *Please machine give apple*, Lana understood the meanings of *Please, machine,* and *give,* let alone the relationships among these symbols that would apply in actual sentences. There is evidence that Sarah and Lana could distinguish the symbol *apple* from symbols that named other reinforcers. This suggests that what Sarah and Lana learned was to produce rote sequences of the type $ABCX$, where A, B, and C are nonsense symbols and X is a meaningful element. The conclusion is supported by the results of two studies, one an analysis of a corpus of Lana's utterances, the other an experiment on serial learning by pigeons.

Thompson and Church (1980) showed that a major portion of the corpus of Lana's utterances can be accounted for by three decision rules that dictate when one of six stock sentences might be combined with one of a small corpus of object or activity names. The decision rules are (1) did Lana want an ingestible object, (2) was the object in view, and (3) was the object in the machine? For example, if the object was in the machine, an appropriate stock sequence was *Please move (object name) into machine*, and so on.

An experiment performed in my laboratory showed that pigeons could learn to peck four colors presented simultaneously in a particular sequence (Straub, Seidenberg, Bever, & Terrace, 1979; Straub & Terrace, 1981). Such performance is of interest as evidence of the memorial capacity of pigeons. It does not, of course, justify interpreting the sequence of the colors $A{\to}B{\to}C{\to}D$ as the production of a sentence meaning *Please machine give grain*.

While the sequences, *Please machine give grain* and *Please machine give apple*, are logically similar, they are not identical. It has yet to be shown that pigeons can learn $ABCX$ sequences of the type that Sarah and Lana learned (where X

stands for different reinforcers) or that pigeons can learn to produce different sequences for different reinforcers. However, given the relative ease with which a pigeon can master an $A \rightarrow B \rightarrow C \rightarrow D$ sequence, neither of these problems seem that difficult a priori. Even if a pigeon could not perform such sequences, or, as would probably be the case, a pigeon learns them more slowly than a chimpanzee, we should not lose sight of the fact that learning a rote sequence does not require any ability to use a grammar.

Utterances of apes who were not explicitly trained to produce rote sequences pose different problems of interpretation. Gardner and Gardner (1974a, 1974b) reported that Washoe was not required to sign sequences of signs nor was she differentially reinforced for particular combinations. She nevertheless signed utterances such as *more drink* and *water bird*. Before these and other utterances can be accepted as creative combinations of signs, that is, combinations that create particular meanings, it is necessary to rule out simpler interpretations.

The simplest nongrammatical interpretation of such utterances is that they contain signs that are related solely by context. Upon being asked what she sees when looking in the direction of a swan it is appropriate for Washoe to sign *water* and *bird*. In this view, if Washoe knew the sign for *sky*, she might just as readily have signed such less interesting combinations as *sky water, bird sky, sky bird water,* and so on.

Even if one could rule out context as the only basis of Washoe's combinations, it remains to be shown that utterances such as *water bird* and *more drink* are constructions in which an adjective and a noun are combined so as to create a new meaning. In order to support that interpretation, it is necessary to show that she combined adjectives and nouns in a particular order. It is, of course, unimportant whether Washoe used the English order (adjective + noun) or the French order (noun + adjective) in creating combinations in which the meaning of a noun is qualified by an adjective. It is important to show, however, that presumed adjectives and nouns are combined in a consistent manner so as to create particular meanings.

It is, of course, true that sign order is but one of many grammatical devices used in sign language. Indeed, sign order is less important in sign language than it is in spoken languages such as English. At the same time, sign order is one of the easiest, if not the easiest, grammatical devices of sign languages to record. It also provides a basis for demonstrating an awareness of such simple constructions as subject-verb, adjective-noun, verb-object, subject-verb-object, and so on.

With only two minor exceptions, Gardner and Gardner have yet to publish any data on sign order as substantiated by a corpus of Washoe's combinations. Accordingly, the interpretation of combinations such as *water bird* and *more drink* remain ambiguous. One has too little information to judge whether such utterances are manifestations of a simple grammatical rule or whether they are merely sequences of contextually related signs.

Project Nim

One way of distinguishing between linguistic and simpler interpretations of an ape's signing is to examine a large body of the ape's utterances for regularities of sign order. The initial goal of Project Nim, a project I started in 1973, was to amass and analyze such a corpus and thereby decide whether a chimpanzee could use one or more rules of finite-state grammar. As puny as such an accomplishment might seem from the perspective of language acquisition by a child, it would amount to a quantal leap in the linguistic ability of nonhumans. As we shall see, showing that a chimpanzee can learn a mere finite-state grammar proved to be an elusive goal.

Socialization and Training

The subject of our study was an infant male chimpanzee, named Nim Chimpsky. Nim was born at the Oklahoma Institute for Primate Studies in November 1973 and was flown to New York at the age of 2 weeks. Until the age of 18 months, he lived in the home of a former student and family; subsequently, he lived in a university-owned mansion in Riverdale, New York, where he was looked after by four students.

At the age of 9 months, Nim became the sole student in a small classroom complex I designed for him in the Psychology Department of Columbia University. The classroom allowed Nim's teachers to focus his attention more easily than they could at home and it also provided good opportunities to introduce Nim to many activities conducive to signing, such as looking at pictures, drawing, and sorting objects. Another important feature of the classroom was the opportunity it provided for observing, filming, and videotaping Nim without his being aware of the presence of visitors and observers who watched him through a one-way window or through cameras mounted in the wall of the classroom.

Nim's teachers kept careful records of what he signed both at home and in the classroom. During each session, the teacher dictated into a cassette recorder as much information as possible about Nim's signing and the context of that signing. Nim was also videotaped at home and in the classroom. A painstaking comparison of Nim's signing in both locales revealed no differences with respect to spontaneity, content, or any other feature of his signing that we examined. In view of comments attributed to other researchers of ape language—that Nim was conditioned in his nursery school like a rat or a pigeon in a Skinner box—it should be emphasized that his teachers were just as playful and spontaneous in the Columbia classroom as they were at home (Bazar, 1980). A detailed description of Nim's socialization and instruction in sign language can be found in *Nim* (Terrace, 1979).

Nim's teachers communicated to him and among themselves in sign language. Although the signs Nim's teachers used were consistent with those of ASL, their signing is best characterized as "pidgin" sign language. This state of affairs is to be expected when a native speaker of English learns sign language. Inevitably, the word order of English superimposes itself on the teacher's signing, at the expense of the many spatial grammatical devices ASL employs. An ideal project would, of course, attempt to use only native or highly fluent signers as teachers —teachers who would communicate exclusively in ASL. There is, however, little reason to be concerned that this ideal has not been realized in Project Nim, or for that matter, in *any* of the other projects that have attempted to teach sign language to apes. The achievements of an ape who truly learned pidgin sign language would be no less impressive than those of an ape who learned pure ASL. Both pidgin sign language and ASL are grammatically structured languages.

Nim was taught to sign by the methods developed by Gardner and Gardner (1969) and Fouts (1972): molding and imitation. During the 44 months he was in New York, he learned 125 signs, most of which were common and proper nouns; next most frequent were verbs and adjectives; least frequent were pronouns and prepositions.

Combinations of Two or More Signs

Lexical Regularities

During a 2-year period, Nim's teachers recorded more than 20,000 of his utterances that consisted of two or more signs. Almost one-half of these utterances were two-sign combinations, of which 1,378 were distinct. One characteristic of Nim's two-sign combinations led me to believe that they were primitive sentences. In many cases Nim used particular signs in either the first or the second position, no matter what other sign that sign was combined with (Terrace, 1979; Terrace, Petitto, Sanders, and Bever, 1979). For example, *more* occurred in the first position in 85% of the two-sign utterances in which *more* appeared (such as *more banana, more drink, more hug,* and *more tickle*). Of the 348 two-sign combinations containing *give*, 78% had *give* in the first position. Of the 946 instances in which a transitive verb (such as *hug, tickle,* and *give*) was combined with *me* and *Nim*, 83% of them had the transitive verb in the first position.

These and other regularities in Nim's two-sign utterances are the first demonstrations I know of a reliable use of sign order by a chimpanzee. By themselves, however, they do not justify the conclusion that they were created according to grammatical rules. Nim could have simply imitated what his teachers were signing. That explanation seemed doubtful for a number of reasons. Nim's teachers had no reason to sign many of the combinations Nim had produced. Nim asked to be tickled long before he showed any interest in tickling; thus, there was no reason for the teacher to sign *tickle me* to Nim. Likewise, Nim requested various objects by signing *give* + *X* (*X* being whatever he wanted) long before be began to offer objects to his teachers. More generally, all of Nim's teachers and many experts on child language learning, some of whom knew sign language, had the clear impression that Nim's utterances typically contained signs that were not imitative of the teacher's signs.

Another explanation of the regularities of Nim's two-sign combinations that does not require the postulation of grammatical competence is statistical. However, an extensive analysis of the regularities observed in Nim's two-sign combinations showed that they did not result from Nim's preferences for using particular signs in the first or second positions of two-sign combinations. Finally, the sheer variety and number of Nim's combinations makes implausible the hypothesis that he somehow memorized them (Terrace, Petitto, Sanders, and Bever, 1980).

Semantic Relationships in Nim's Two-Sign Combinations

Semantic distributions, unlike the lexical ones discussed above, cannot be constructed directly from a corpus. In order to derive a semantic distribution, observers have to make judgments as to what each combination means. Procedures for making such judgments, introduced by Bloom (1970, 1973; Bloom & Lahey, 1971) and Schlesinger (1971) are known as the method of *rich interpretation* (see also Brown, 1973; Fodor, Bever, & Garrett, 1974). An observer relates the utterance's immediate context to its contents. Supporting evidence for semantic judgments includes the following observations. The child's choice of word order is usually the same as it would be if the idea were being expressed in the canonical adult form. As the child's mean length of utterance (MLU) increases, semantic relationships identified by a rich interpretation develop in an orderly fashion (Bloom, 1973; Bowerman, 1973; Brown, 1973). The relationships expressed in two-word combinations are the first ones to appear in the three- and four-word combinations. Many longer utterances appear to be composites of the semantic relationships expressed in shorter utterances (Bloom, 1973; Brown, 1973).

Studies by Gardner and Gardner and by Patterson of an ape's ability to express semantic relationships in combinations of signs have yet to advance beyond the stage of unvalidated interpretation. Gardner and Gardner (1971) and Patterson (1978) concluded that a substantial portion of Washoe's and Koko's two-sign combinations were interpretable in categories similar to those used to describe two-word utterances of children (78% and 95% of Washoe's and Koko's two-sign combinations, respectively). No data are available as to the reliability of the interpretations that Gardner and Gardner and Patterson have advanced.

Without predjudging whether Nim actually expressed semantic relationships in his combinations, we analyzed, by the method of rich interpretation, 1,262 of his two-sign combinations, which occurred between the ages of 25 and 31 months. Twenty categories of semantic relationships account for 895 (85%) of the 957 interpretable two-sign combinations. Brown (1973) found that 11 semantic relationships account for about 75% of all combinations of the children he studied. Similar categories of semantic relationships were used by Gardner and Gardner (1971) and by Patterson (1978).

Nim showed significant preferences for placing signs expressing a particular semantic role in either the first or the second positions (Figure 1-1). Agent, at-

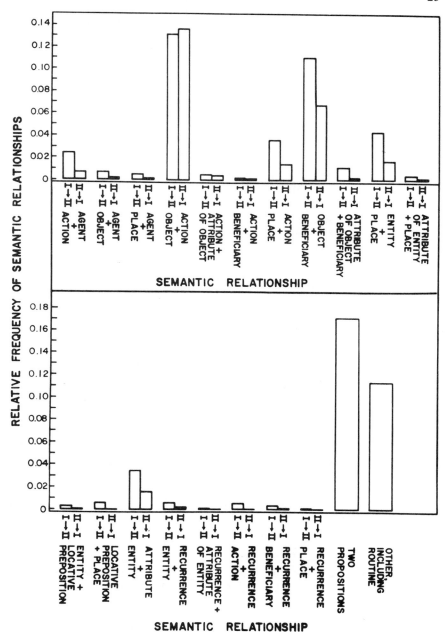

Figure 1-1. Relative frequencies of different semantic relationships. The bars above I and II show the relative frequencies of two-sign combinations expressing the relationship in the order specified under the bar, for example, an agent followed by an action. The bars above II and I show the relative frequencies of two-sign combinations expressing the same relationship in the reverse order, for example, an action followed by an agent.

tribute, and recurrence (*more*) were expressed most frequently by second-position signs.[1]

Differences Between Nim's and a Child's Combinations of Signs and Words

The analyses performed on Nim's combinations provided the most compelling evidence I know of that a chimpanzee could use grammatical rules, albeit finite-state rules, for generating two-sign sequences. It was not until after our funds ran out and it became necessary to return Nim to the Oklahoma Institute for Primate Studies that I became skeptical of that conclusion. Ironically, it was our newly found freedom from data collecting, teaching, and looking after Nim that allowed me and other members of the project to examine Nim's use of sign language more thoroughly. What emerged from our new analyses was a number of important differences between Nim's and a child's use of language. One of the first facts that troubled me was the absence of any increase in the length of Nim's utterances. During the last 2 years that Nim was in New York, the average length of Nim's utterances fluctuated between 1.1 and 1.6 signs. That performance is similar to what children do when they begin combining words. Furthermore, the maximum length of a child's utterances is related very reliably to their average length. Nim's showed no such relationship.

As children get older, the average length of their utterances increases steadily. As shown in Figure 1-2, this is true both of children with normal hearing and of deaf children who sign. After learning to make utterances relating a verb and an object (e.g., *eats breakfast*) and utterances relating a subject and a verb (e.g., *Daddy eats*), the child learns to link them into longer utterances relating the subject, verb, and object (e.g., *Daddy eats breakfast*). Later, the child learns to link them into longer utterances such as *Daddy didn't eat breakfast,* or, *When will Daddy eat breakfast?*

Despite the steady increase in the size of Nim's vocabulary, the mean length of his utterances did not increase. Although some of his utterances were very long, they were not very informative. Consider, for example, his longest utterance, which contained 16 signs: *give orange me give eat orange me eat orange give me eat orange give me you.* The same kinds of run-on sequences can be seen in comparing Nim's two-, three-, and four-sign combinations. As shown in Table 1-1, the topic of Nim's three-sign combinations overlapped considerably with the apparent topic of his two-sign combinations. Eighteen of Nim's 25 most fre-

---

[1] The relative frequency of combinations discussed on the preceding pages is based on exhaustive lexical analysis of such combinations as derived from the total corpus of Nim's combinations. The data shown in Figure 1-1 refer to a subset of that corpus for which contextual information was available (from videotapes and/or teacher's notes). This contextual information provided a basis for making judgments about the meanings of the combinations summarized in Figure 1-1.

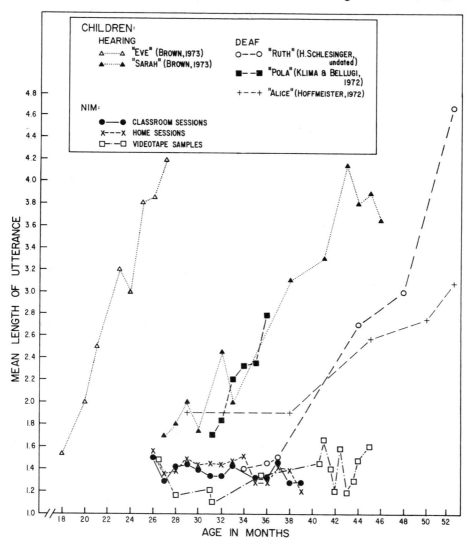

Figure 1-2. Mean length of signed utterances of Nim and three deaf children and mean length of spoken utterances of two hearing children. The functions showing Nim's MLU between 29 and 39 months of age are based on data obtained from teachers' reports; the function showing Nim's MLU between 27 and 45 months of age is based upon videotape transcript data.

quent two-sign combinations can be seen in his 25 most frequent three-sign combinations, in virtually the same order in which they appear in his two-sign combinations. Furthermore, if one ignores sign order, all but five signs that appear in Nim's 25 most frequent two-sign combinations—*gum, tea, sorry, in,* and *pants*—appear in his 25 most frequent three-sign combinations. We did not have enough

Table 1-1. Twenty-five Most Frequent Two- and Three-Sign Combinations

Two-sign combinations	Frequency	Three-sign combinations	Frequency
play me	375	play me Nim	81
me Nim	328	eat me Nim	48
tickle me	316	eat Nim eat	46
eat Nim	302	tickle me Nim	44
more eat	287	grape eat Nim	37
me eat	237	banana Nim eat	33
Nim eat	209	Nim me eat	27
finish hug	187	banana eat Nim	26
drink Nim	143	eat me eat	22
more tickle	136	me Nim eat	21
sorry hug	123	hug me Nim	20
tickle Nim	107	yogurt Nim eat	20
hug Nim	106	me more eat	19
more drink	99	more eat Nim	19
eat drink	98	finish hug Nim	18
banana me	97	banana me eat	17
Nim me	89	Nim eat Nim	17
sweet Nim	85	tickle me tickle	17
me play	81	apple me eat	15
gum eat	79	eat Nim me	15
tea drink	77	give me eat	15
grape eat	74	nut Nim nut	15
hug me	74	drink me Nim	14
banana Nim	73	hug Nim hug	14
in pants	70	play me play	14
		sweet Nim sweet	14

contextual information to perform a semantic analysis of Nim's two- and three-sign combinations. However, Nim's teachers' reports indicate that the individual signs of his combinations were appropriate to their context and that equivalent two- and three-sign combinations occurred in the same context.

Although lexically similar to two-sign combinations, the three-sign combinations do not appear to be informative elaborations of two-sign combinations. Consider, for example, Nim's most frequent two- and three-sign combinations: *play me* and *play me Nim*. Combining *Nim* with *play me* to produce the three-sign combination, *play me Nim*, adds a redundant proper noun to a personal pronoun. Repetition is another characteristic of Nim's three-sign combinations, for example, *eat Nim eat*, and *nut Nim nut*. In producing a three-sign combination, it appears as if Nim was adding to what he might sign in a two-sign combination, not so much to add new information, but instead to add emphasis. Nim's most frequent four-sign combinations reveal a similar picture. In children's utterances, by contrast, the repetition of a word, or a sequence of words, is a rare event.

Having seen what Nim signed about in two-, three-, and four-sign combinations, it is instructive to see what he signed about with single-signs. As shown in Table 1-2, the topics of Nim's most frequent 25 single-sign utterances overlap considerably with those of his most frequent multisign utterances. (Most of the exceptions are signs required in certain routines, e.g., *finish*, when Nim was finished using the toilet; *sorry*, when Nim was scolded, and so on.) In contrast to a child's longer utterances which are semantically and syntactically more complex than his shorter utterances, Nim's are not. When signing a combination, as opposed to signing a single sign, Nim gave the impression that he was running on with his hands. It appears that Nim learned that the more he signed, the better his chances were for obtaining what he wanted. It also appeared as if Nim made no effort to add informative, as opposed to redundant, signs in satisfying his teacher's demand that he sign.

At first glance, the results of our semantic analysis appear to be consistent with the observations of Gardner and Gardner and those of Patterson. Nevertheless, even though I could demonstrate the reliability of our judgments, several features of our results suggest that our analysis, and that of others, may exagger-

Table 1-2. Twenty-five Most Frequent Single Signs (July 5, 1976 to February 7, 1977)

Sign	Tokens
hug	1,650
play	1,545
finish	1,103
eat	951
dirty	788
drink	712
out	615
Nim	613
open	554
tickle	414
bite	407
shoe	405
red	380
pants	372
sorry	366
angry	354
me	351
banana	348
nut	323
down	316
toothbrush	302
more	301
grape	239
sweet	236

ate an ape's semantic competence. One problem is the subjective nature of semantic interpretations. That problem can be remedied only to the extent that evidence corroborating the psychological reality of our interpretations is available. Neither our semantic analyses of Nim's two-sign combinations nor those of any other studies have produced such evidence. In some cases, utterances were inherently equivocal in our records. Accordingly, somewhat arbitrary rules were used to interpret these utterances. Consider, for example, combinations of *Nim* and *me* with an object name (for example, *Nim banana*). These occurred when the teacher held up an object that the teacher was about to give to Nim who, in turn, would ingest it. We had no clear basis for distinguishing between the following semantic interpretations of combinations containing *Nim* or *me* and an object name: agent–object, beneficiary–object, and possessor–possessed object.

An equally serious problem is posed by the very small number of lexical items used to express particular semantic roles. Only when a semantic role is represented by a large variety of signs is it reasonable to attribute position preferences to semantic rules rather than to lexical position habits. For example, the role of recurrence was presented exclusively by *more*. In combinations presumed to relate an agent and an object or an object and a beneficiary, one would expect agents and beneficiaries to be expressed by a broad range of agents and beneficiaries, for example: *Nim, me, you,* and names of other animate beings. However, 99% (N=297) of the beneficiaries in utterances judged to be object–beneficiary combinations were *Nim* and *me*, and 76% (N=35) of the agents in utterances judged to be agent–object combinations were *you*. From these and other examples, it is difficult to decide whether the positional regularities favoring agent–object and object–beneficiary constructions (Figure 1-1) are expressions of semantic relationships or idiosyncratic lexical position habits. Such isolated effects could also be expected to appear from statistically random variation.

The most dramatic difference between Nim's and a child's use of language was revealed in a painstaking analysis of videotapes of Nim's and his teacher's signing. These tapes revealed much about the nature of Nim's signing that could not be seen with the naked eye. Indeed, they were so rich in information that it took as much as 1 hour to transcribe a single minute of tape.

Nim's signing with his teachers bore only a superficial resemblance to a child's conversations with his or her parents (Terrace et al., 1979). What is more, only 12% of Nim's utterances were not preceded by a teacher's utterance. A significantly larger proportion of a child's utterances is spontaneous.

In addition to differences in spontaneity, there were differences in creativity. As a child gets older, the proportion of utterances that are full or partial imitations of the parent's prior utterance(s) decreases from less than 20% at 21 months to almost zero by the time the child is 3 years old. When Nim was 26 months old, 38% of his utterances were full or partial imitations of his teacher's. By the time he was 44 months old, the proportion had risen to 54%. Sanders (1980) showed that the function of Nim's imitative utterances differed from those of a child's imitative utterances. Bloom and her associates (1975) observed that children imitate their parents' utterances mainly when they were learning new words or new syntactic structures. Sanders found no evidence for either of these imitative functions in Nim's imitative utterances.

As children imitate fewer of their parents' utterances, they begin to expand upon what they hear their parents say. At 21 months, 22% of a child's utterances add at least one word to the parent's prior utterance; at 36 months, 42% are expansions of the parent's prior utterance. Fewer than 10% of Nim's utterances recorded during 22 months of videotaping (the last 22 months of the project) were expansions. Like the mean length of his utterances, this value remained fairly constant.

The videotapes showed another distinctive feature of Nim's conversations that we could not see with the naked eye. He was as likely to interrupt his teacher's signing as not. In contrast, children interrupt their parents so rarely (so long as no other speakers are present) that interruptions are all but ignored in studies of their language development. A child learns readily what one takes for granted in a two-way conversation: each speaker adds information to the preceding utterance and each speaker takes a turn in holding the floor. Nim rarely added information and showed no evidence of turn-taking.

None of the features of Nim's discourse—his lack of spontaneity, his partial imitation of his teacher's signing, his tendency to interrupt—had been noticed by any of his teachers or by the many expert observers who had watched Nim sign. Once I was sure that Nim wasn't imitating *precisely* what his teacher had just signed, I felt that it was less important to record the teachers' signs than it was to capture as much as I could about Nim's signing: the context and specific physical movements, what hand he signed with, the order of his signs, and their appropriateness. Even if one wanted to record the teacher's signs, limitations of attention span would make it too difficult to remember all of the significant features of *both* the teacher's and Nim's signs.

Once I knew what to look for, the contribution of the teacher was easy to see, embarrassingly enough in still photographs that I had looked at for years. Consider, for example, Nim's signing the sequence *me hug cat* as shown in Figure 1-3. At first, these photographs (and many others) seemed to provide clear examples of spontaneous and meaningful combinations; but just as analyses of our videotapes provided evidence of a relationship between Nim's signing and the teacher's prior signing, a reexamination of Figure 1-3 revealed a previously unnoticed contribution of the teacher's signing. She signed *you* while Nim was signing *me* and signed *who* while he was signing *cat*. While Nim was signing *hug*, his teacher held her hand in the *n*-hand configuration, a prompt for the sign *Nim*. Because these were the only photographs taken of this sequence, we cannot specify just when the teacher began her signs. It is not clear, for example, whether the teacher signed *you* simultaneously or immediately prior to Nim's *me*. It is, however, unlikely that the teacher signed *who* after Nim signed *cat*. A few moments before these photographs were taken the teacher repeatedly quizzed Nim as to the contents of the cat box by signing *who*. At the very least, Nim's sequence, *me hug cat*, cannot be interpreted as a spontaneous combination of three signs.

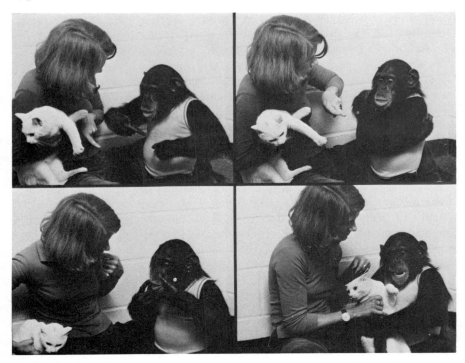

Figure 1-3. Nim signing the linear combination, *me hug cat* to his teacher, Susan Quinby (Photographed in classroom by H. S. Terrace.)

There remain other differences between Nim's and a child's signing that need to be explored. Two important questions are: How similar are the contents of a chimpanzee's and a child's utterances? Are the utterances structured grammatically? A cursory examination of corpora of children's utterances indicates that chimpanzees and children differ markedly with respect to the variety of the utterances they make. While children produce certain routine combinations, they are relatively infrequent. Nim's utterances, on the other hand, show a high frequency of routine combinations (e.g., *play me, me Nim*; see Table 1-1 for additional examples).

It also remains to be shown that the patterns of vocal discourse observed between a hearing parent and a hearing child are similar to the patterns of signing discourse that obtain between a deaf parent and a deaf child. Videotapes of such discourse are now available from a number of sources (Bellugi & Klima, 1976; Hoffmeister, 1978; McIntire, 1978).

Criticisms of Project Nim

Of more immediate concern is the generality of the conclusions drawn about Nim's signing. This issue can be approached in two ways. One is to consider the methodological weaknesses of Project Nim and to pursue their implications. The other is to ignore Project Nim and to ask whether other signing apes sign because they are coaxed to do so by the teachers and how much overlap exists between the teacher's and the ape's signing.

Consider first some of the initial questions raised about Project Nim. It has been said that Nim was taught by too many teachers (60 all told), that his teachers were not fluent enough in ASL, that terminating Nim's training at the age of 44 months prevented his teachers from developing Nim's full linguistic competence, and that Nim may simply have been a dumb chimp. (See Bazar, 1980, for references to other researchers studying language in apes.)

Aside from the speculation that Nim may have been a dumb chimp, I believe that all of these criticisms are valid; and, if questioning Nim's intelligence is simply a nasty way of asking whether an N of 1 is an adequate basis for forming a general conclusion about an ape's grammatical competence, I would readily admit that it is not. A case can be made, however, that most of the methodological inadequacies of Project Nim have been exaggerated and, in any event, that they are hardly unique to Project Nim. Although Nim was taught by 60 teachers he spent most of his time in the presence of a core group of 8 teachers: Stephanie LaFarge, Laura Petitto, Amy Schachter, Walter Benesch, Bill Tynan, Joyce Butler, Richard Sanders, and myself. As described in *Nim* (Terrace, 1979; see especially Appendix B), many of Nim's 60 teachers served as occasional playmates rather than as regular teachers. Nevertheless, they were each listed as a teacher. Gardner and Gardner have yet to publish a full list of the teachers who worked with Washoe. Fouts (1980), however, estimates that their number was approximately 40. Of greater importance is Fouts' observation that, during Washoe's 4 years in Reno, Nevada, she was looked after mainly by a small group of 6 core teachers.

Patterson (1979) presented data provided by 20 teachers during the 3-year period in which they worked with Koko. This large number of teachers is hardly surprising. It was difficult for an ape's teachers to sustain the energy needed to carry out lesson plans, to engage its attention, to stimulate it to sign, and to record what it signed for more than 3–4 hours/day. A 3- to 4-hour session with Nim also entailed an additional 1 or 2 hours to transcribe the audio cassette on which the teacher dictated information about Nim's signing and to write a report At least six full-time people would be needed to carry out such a schedule on a 16-hour/day, 7-day/week basis. As far as I know, no project has been able to afford the salaries of such a staff. Accordingly, it is necessary to make do with a large contingent of part-time volunteers.

Both Patterson and Fouts speak English while signing with their apes. Films of Gardner and Gardner and Fouts signing with Washoe, of Fouts signing with Ally and Booee (two resident chimpanzees of the Oklahoma Institute for Primate Studies), and of Patterson signing with Koko make it clear that none of these researchers use ASL. As mentioned earlier, pidgin sign language seems to be the prevalent form of communication on *all* projects attempting to teach apes to use sign language.

Washoe is now 15, Koko is 9, and Ally is 9 years old. I know of no evidence that their linguistic skills increased as they became older. An ape's intelligence undoubtedly increases after infancy. One must, however, also keep in mind that as an ape gets older, its ability to master its environment by physical means also increases. An ape's increasing strength and its recognition that it can get its way without signing would make the teacher less dominant. As a result the ape would be *less* motivated to sign. I am therefore skeptical of conjectures that an ape's increasing intelligence would manifest itself in a more sophisticated use of language.[2]

Whatever the shortcomings of Project Nim, it should be recognized that they are irrelevant to the following hypothesis about an ape's use of signs: An ape signs mainly in response to his teacher's urgings, in order to obtain certain objects or activities. Combinations of signs are not used creatively to generate particular meanings. Instead, they are used for emphasis or in response to the teacher's unwitting demands that the ape produce as many contextually relevant signs as possible. The validity of this hypothesis rests simply on the nature of data obtained from other signing apes.

Finding such data has proven difficult, particularly because discourse analyses of other signing apes have yet to be published. Furthermore, as mentioned earlier, published accounts of an ape's combinations of signs have centered around anecdotes and not around exhaustive listings of all combinations. One can, however, obtain some insight into the nature of signing by other apes by looking at films and videotape transcripts of their signing. Two films made by Gardner and Gardner (1973, 1976) of Washoe's signing, a doctoral dissertation by Miles (1978; it contains four videotape transcripts of two Oklahoma chimps, Ally and Booee), and a recently released film *Koko, A Talking Gorilla* (Schroeder, 1973), all support the hypothesis that the teacher's coaxing and cueing have played much greater roles in so-called conversations with chimpanzees than was previously recognized.

In a film produced by *Nova* (1976), B. T. Gardner can be seen signing *what time now?*, an utterance Washoe interrupts to sign, *time eat, time eat*. A longer version of the same exchange shown in the second film began with B. T. Gardner signing *eat me, more me*, after which Washoe gave Gardner something to eat (Gardner & Gardner, 1976). Then she signed *thank you*—and asked *what time now?* Washoe's response *time eat, time eat* can hardly be considered spontaneous, since Gardner had just used the same signs and Washoe was offering a direct answer to her question.

The potential for misinterpreting an ape's signing because of inadequate reporting is made plain by another example in both films. Washoe is conversing with her teacher, Susan Nichols, who shows the chimp a tiny doll in a cup.

---

[2] Two recent reviews of *Nim* (Terrace, 1979), one by B. T. Gardner (1981), and one by Gaustad (1981), a graduate student of the Gardens, have raised additional objections about the conduct of Project Nim and my conclusions regarding the nature of ape signing. These critiques make patently false claims, for instance, that Nim was not taught to sign in his "Spartan" classroom. At best these reviews present facts stripped of their context; at worst they simply propagate misinformation. A full discussion of these issues can be found in my replies to Gardner and Gaustad (Terrace, 1981, 1982).

Nichols points to the cup and signs *that*; Washoe signs *baby*. Nichols brings the cup and doll closer to Washoe, allowing her to touch them, slowly pulls them away, and then signs *that* by pointing to the cup. Washoe signs *in* and looks away. Nichols brings the cup and doll closer to Washoe again, who looks at the two objects once more, and signs *baby*. Then, as she brings the cup still closer, Washoe signs *in*. *That*, signs Nichols, and points to the cup; *my drink*, signs Washoe.

Given these facts, there is no basis to refer to Washoe's utterance—*baby in baby in my drink*—as a spontaneous or creative use of *in* as a preposition joining two objects. It is actually a run-on sequence with very little relationship between its parts. Only the last two signs were uttered without prompting from the teacher. Moreover, the sequence of the prompts (pointing to the doll, and then pointing to the cup) follows the order called for in constructing an English prepositional phrase. In short, discourse analysis makes Washoe's achievement less remarkable than it might seem at first.

In commenting about the fact that Koko's MLU was low in comparison with that of both hearing and deaf children, Patterson (1979) speculated that "this probably reflects a species difference in syntactic and/or sequential processing abilities." Patterson went on to observe that "the majority of Koko's utterances were not spontaneous, but elicited by questions from her teachers and companions. My interactions with Koko were often characterized by frequent questions such as 'What's this?'."(p. 153).

Four transcripts appended to Miles' (1978) dissertation provided me with a basis for performing a discourse analysis of the signing of two other chimpanzees, Ally and Booee. Each transcript presents an exhaustive account of one of these chimps signing with one of two trainers: Roger Fouts and Joe Couch. The MLU and a summary of the discourse analysis of each tape are shown in Table 1-3. In Tables 1-4 and 1-5 exhaustive summaries of two conversations are shown: one from a session with the highest MLU, and one from a session with the highest percentage of adjacent utterances that were novel. The novel utterances are very similar to Nim's run-on sequences. They also overlap considerably with adjacent utterances that were expansions and with noncontingent utterances.

Table 1-3. Summary of Ally and Booee Transcripts

Parameter	Videotape no.				
	3	4	5	6	Mean
No. of utterances	38	79	102	72	72.75
MLU	1.63	1.52	2.25	1.93	1.85
Adjacent (%)	76.3	93.7	77.4	86.1	83.4
Imitations (%)	13.6	22.8	7.84	8.33	13.03
Expansions (%)	7.89	7.59	13.7	4.16	8.34
Novel (%)	55.3	63.3	55.9	73.6	62.3
Noncontingent (%)	23.7	6.3	2.6	3.9	16.6

Table 1-4. Summary of "Conversation" Between Roger Fouts and Ally

Sign	Frequency
Adjacent Utterances ($N=74$; $C=67$; ?=7)	
Novel ($N=50$; $C=43$; ?=7)	
Roger	14
Roger tickle Ally	9
Roger tickle	3
tickle Roger	1
tickle	1
Roger tickle Ally hurry	1
George	2
Joe	7
string	1
that	2
that that box	1
that shoe	1
Roger string George	1
comb	1
good	1
food-eat	2
Ally	2
Expansions ($N=6$; $C=6$)	
George smell Roger	1
tickle hurry	1
Roger tickle	1
Roger tickle Ally	1
Roger tickle Roger comb	1
Roger tickle Roger	1
Imitations ($N=18$; $C=18$)	
Roger	5
tickle	2
Roger tickle Ally	1
shoe tickle	2
baby	1
pillow	1
comb	2
pull	2
Nonadjacent Utterances ($N=5$)	
Joe	2
you tickle	1
shoe	1
pillow	1

Abbreviations: N, total number of particular type of utterance; C, utterances that followed a teacher's command; ?, utterances that followed a teacher's question.

Table 1-5. Summary of "Conversation" Between Joe Couch and Booee

Sign	Frequency
Adjacent Utterances (N=79; C=31; ?=47)	
Novel (N=57; C=21; ?=35)	
food-eat Booee that	3
food-eat Booee	3
food-eat	3
you food-eat Booee that	1
Booee food-eat more Booee	1
food-eat Booee hungry	1
food-eat me fruit Booee	1
that food-eat Booee	1
fruit food-eat you	1
food-eat fruit hurry	1
gimme food-eat	1
gimme	1
fruit more Booee	1
gimme fruit hurry	1
there more fruit	1
hurry	4
fruit	1
that Booee over there	1
that	4
more that Booee	1
that Booee	1
more Booee	1
Booee	2
you Booee	1
more	1
you	3
Booee hurry	1
there you gimme	1
you Booee you Booee hurry	1
more Booee hurry	1
tickle Booee	3
tickle you me	1
you there	1
hurry tickle Booee hurry	1
you there that there	1
that Booee there Booee	1
baby that	1
more baby	1
Expansions (N=14; C=4; ?=10)	
hurry tickle Booee hurry Booee hurry	1
Booee you Booee hurry gimme	1
tickle Booee gimme	1
me tickle	1

Table 1-5. (*Continued*)

Sign	Frequency
me tickle hurry	1
food eat Booee that	1
that Booee	1
that tickle	1
there you	1
Imitation (*N*=8; *C*=6; ?=2)	
that	4
you	2
tickle	1
baby	1
Nonadjacent Utterances (*N*=23)	
tickle Booee	6
tickle	2
more Booee	1
hurry that tickle	1
tickle gimme	1
food-eat	1
food-eat Booee	1
more fruit there	1
that	1
that that Booee gimme	1
gimme	2
Booee	1
hurry gimme hurry	1
that there	1
that more baby	1
there that there	1

Abbreviations: *N*, total number of particular type of utterance; *C*, utterance following a teacher's command; ?, utterances following a teacher's question.

In his discussion of communicating with an animal, the philosopher Wittgenstein (1963) cautioned that apparent instances of an animal using human language may prove to be a "game" that is played by simpler rules. Nim's, Washoe's, Ally's, Booee's, and Koko's use of signs suggests a type of interaction between an ape and its trainer that has little to do with human language. In each instance the sole function of the ape's signing appears to be to request various rewards that can be obtained only by signing. Little, if any, evidence is available that an ape signs in order to exchange information with its trainer, as opposed to simply demanding some object or activity.

In a typical exchange the teacher first tries to interest the ape in some object or activity such as looking at a picture book, drawing, or playing catch. Typically the ape tries to engage in such activities without signing. The teacher then tries to initiate signing by asking questions such as *What that? What you want? who's book? Ball red or blue?* The more rapidly the ape signs, the more rapidly it can obtain what it wants. It is therefore not surprising that the ape frequently interrupts the teacher. From the ape's point of view, the teacher's signs provide an excellent model of the signs it is expected to make. By simply imitating a few of them, often in the same order used by the teacher, and by adding a few "wild cards"—general purpose signs such as *give, me, Nim,* or *more*—the ape can produce utterances that appear to follow grammatical rules. What seems like conversation from a human point of view is actually an attempt to communicate a demand (in a nonconversational manner) as quickly as possible.

Future Research

It might be argued that signing apes have the potential to create sentences but did not do so because of motivational rather than intellectual limitations. Perhaps Nim and Washoe would have been more motivated to communicate in sign language if they had been raised by smaller and more consistent groups of teachers, thus sparing them emotional upheavals. It is, of course, possible that a new project, administered by a permanent group of teachers who are fluent in sign language and have the skills necessary for such experiments, would prove successful in getting apes to create sentences.

It is equally important for any new project to pay greater attention to the function of the signs than to mastery of syntax. In the rush to demonstrate grammatical competence in the ape, many projects (Project Nim included) overlooked functions of individual signs other than their demand function. Of greater significance, from a human point of view, are the abilities to use a word simply to communicate information and to refer to things which are not present. One would like to see, for example, to what extent an ape is content to sign *flower* simply to draw the teacher's attention to a flower with no expectation that the teacher would give it a flower. In addition one would want to see whether an ape could refer symbolically to objects that are not in view in order to exchange information about those objects. For example, could an ape respond, in a nonrote manner, to a question such as *What color is the banana?* by signing *yellow* or to a question such as *Who did you chase before?* by signing *cat.* Until it is possible to teach an ape that signs can convey information other than mere demands it is not clear why an ape would learn a grammatical rule. To put the question more simply, why should an ape be interested in learning rules about relationships between signs when it can express all it cares to express through individual signs?

The personnel of a new project would have to be on guard against the subtle and complex imitation that was demonstrated in Project Nim. In view of the discoveries about the nature of Nim's signing that were made through videotape analyses, it is essential for any new project to maintain a permanent and unedited visual record of the ape's discourse with its teachers. Indeed, the absence of such documentation would make it impossible to substantiate any claims concerning the spontaneity and novelty of an ape's signing.

Requiring proof that an ape is not just mirroring the signs of its teachers is not unreasonable. Indeed, it is essential for any researcher who seeks to determine, once and for all, whether apes can use language in a human manner. It is unreasonable to expect that in any such experiment ape "language" must be measured against a child's sophisticated ability; that ability still stands as an important definition of the human species.

While writing "A Report to an Academy," Kafka obviously had no way of anticipating the numerous attempts to teach real apes to talk that took place in the United States and in the Soviet Union (Fouts, 1972; Gardner & Gardner, 1969; Hayes, 1951; Kellogg & Kellogg, 1933; Kellogg, 1968; Kots, 1935; Premack, 1970; Rumbaugh et al., 1973; Temerlin, 1975; Terrace et al., 1979). Just the same, his view that an ape will imitate for "a way out" seems remarkably telling. If one substitutes for the phrase, "a way out," rewarding activities such as being tickled, chased, hugged, access to a pet, cat, books, drawing materials, and items of food and drink, the basis of Nim's, Washoe's, Koko's and other apes' signing seems adequately explained. Much as I would have preferred otherwise, a chimpanzee's "Report to an Academy" remains a work of fiction.

Acknowledgments

Portions of this chapter appeared previously in "How Nim Chimpsky Changed My Mind," *Psychology Today*, November, 1979, 13, 65–76. This work was supported in part by grants from the W. T. Grant Foundation, the Harry Frank Guggenheim Foundation, and the National Institutes of Health (Grant RO1 MH29293).

References

Bazar, J. Catching up with the ape language debate. *American Psychological Association Mentor*, 1980, *11*, 4–5, 47.

Bellugi, U., & Klima, E. S. *The signs of language*. Cambridge, Mass.: Harvard University Press, 1976.

Bloom, L. M. *Language development: Form and function in emerging grammars*. Cambridge, Mass.: MIT Press, 1970.

Bloom, L. M. *One word at a time: The use of single word utterances before syntax*. The Hague: Mouton, 1973.

Bloom, L. M., & Lahey, M. *Language development and language disorders*. New York: Wiley, 1971.

Bloom, L. M., Lightbrown, P., & Hood, L. Structure and variation in child language. In *Monographs of the Society for Research in Child Development*, vol. 40 (1975) ser. 160.

Bowerman, M. Structural relationships in children's utterances: Syntactic or semantic? In *Cognitive development and acquisition of language*. T. E. Moore (Ed.), New York: Academic Press, 1973.

Brown, R. *A first language: The early stage*. Cambridge, Mass.: Harvard University Press, 1973.

Chomsky, N. *Syntactic structures*. The Hague: Mouton, 1957.

Fodor, J. A., Bever, T. G., & Garrett, M. F. *The psycology of language: An introduction to psycholinguistics and generative grammar*. New York: McGraw-Hill, 1974.

Fouts, R. S. Use of guidance in teaching sign language to a chimpanzee (Pan troglodytes). *Journal of Comparative and Physiological Pyschology*, 1972, *80*, 515–522.

Fouts, R. S. Personal communication, March 1980.

Gardner, B. T. Project Nim: Who taught whom? *Contemporary Psychology* 1981, *26*, 425–427.

Gardner, B. T., & Gardner, R. A. Two-way communication with an infant chimpanzee. In A. M. Schrier & F. Stollnitz (Eds.), *Behavior of nonhuman primates* (Vol. 4). New York: Academic Press, 1971.

Gardner, B. T., & Gardner, R. A. Comparing the early utterances of child and chimpanzee. In A. Pick (Ed.), *Minnesota symposia on child psychology* (Vol. 8). Minneapolis: University of Minnesota Press, 1974. (a)

Gardner, B. T., & Gardner, R. A. Teaching sign language to a chimpanzee, VII: Use of order in sign combinations. *Bulletin of the Psychonomic Society*, 1974, *4*, 264–267. (b)

Gardner, R. A., & Gardner, B. T. Teaching sign language to a chimpanzee. *Science*, 1969, *165*, 664-672.

Gardner, R. A., & Gardner, B. T. (Producers). *Teaching sign language to the chimpanzee Washoe*. University Park, PA: Psychological Cinema Register, 1973. No. 16802. (Film)

Gardner, R. A., & Gardner, B. T. (Producers). *The first signs of Washoe*. New York: Time-Life Films/WGBH-Nova, 1976. (Film)

Gaustad, G. R. Review of *Nim. Sign Language Studies*, 1981, *30*, 89–94.

Hayes, C. *The ape in our house*. New York: Harper & Row, 1951.

Hoffmeister, R. J. *The development of demonstrative pronouns, locatives and personal pronouns in the acquisition of American Sign Language by deaf children of deaf parents*. Unpublished doctoral dissertation, University of Minnesota, 1978.

Kafka, F. *Selected short stories by Franz Kafka* (W. Muir & E. Muir, trans.). New York: Modern Library, 1952. (Originally published, 1917.)

Kellogg, L. A., & Kellogg, W. N. *The ape and the child: A study of environmental influence and its behavior*. New York: McGraw-Hill, 1933.

Kellogg, W. N. Communication and language in home-raised chimpanzees. *Science*, 1968, *182*, 423–427.

Klima, E. S., & Bellugi, U. The signs of language in child and chimpanzee. In T. Alloway, L. Kramer, & P. Pliner (Eds.), *Communications and affect: A comparative approach*. New York: Academic Press, 1972.

Kots, N. *Infant ape and human child*. Moscow: Museum Darwinianum, 1935.

McIntire, M. L. *Learning to take your turn in ASL* (Working paper). Unpublished manuscript, Department of Linguistics, University of California, Los Angeles, 1978.

Miles, H. L. *Conversations with apes: The use of sign language by two chimpanzees*. Unpublished doctoral dissertation, University of Connecticut, 1978.

Miller, G. A. The psycholinguists. *Encounter,* 1964, *23,* 29–37.

Patterson, F. G. The gestures of a gorilla: Language acquisition by another pongid. *Brain and Language,* 1978, *12,* 72–97.

Patterson, F. G. *Linguistic capabilities of a lowland gorilla.* Unpublished doctoral dissertation, Stanford University, 1979.

Premack, D. A functional analysis of language. *Journal of Experimental Analysis of Behavior,* 1970, *4,* 107–125.

Rumbaugh, D. M., Gill, T. V., & Glaserfeld, E. C. von. Reading and sentence completion by a chimpanzee *(Pan). Science,* 1973, *192,* 731–733.

Sanders, R. J. *The influence of verbal and nonverbal context of the sign language conversations of a chimpanzee.* Unpublished doctoral disseration, Columbia University, 1980.

Schlesinger, H. Unpublished manuscript, 1976.

Schlesinger, I. N. Production of utterances and language acquisition. In D. I. Slobin (Ed.), *Ontogenesis of grammar.* New York: Academic Press, 1971.

Schroeder, B. (Producer). *Koko, a talking gorilla.* New Yorker Films, 1973. (Film)

Straub, R. L., Seidenberg, M. S., Bever, T. G., & Terrace, H. S. Serial learning in the pigeon. *Journal of Experimental Analysis of Behavior,* 1979, *32,* 137–148.

Straub, R. L., & Terrace, H. S. Generalization of serial learning in the pigeon. *Animal Learning and Behavior,* 1981, *9,* 454–468.

Temerlin, M. K. *Lucy: Growing up human: A chimpanzee daughter in a psychotherapist's family.* Palo Alto, Calif.: Science and Behavior, 1975.

Terrace, H. S. *Nim.* New York: Knopf, 1979.

Terrace, H. S. Evidence for sign language in apes: What the ape signed, or how well was the ape loved? *Contemporary Psychology,* 1981, *27,* 67–68.

Terrace, H. S. Language in apes: Fact or projection? *Sign Language Studies,* 1982, *35,* 178–180.

Terrace, H. S., Petitto, L. A., Sanders, R. J., & Bever, T. G. Can an ape create a sentence? *Science,* 1979, *206,* 891-902.

Terrace, H. S., Petitto, L. A., Sanders, R. J., & Bever, T. G. On the grammatical capacity of apes. In K. Nelson (Ed.) *Children's Language.* New York: Gardner Press, 1980.

Thompson, C. R., & Church, R. M. An explanation of the language of a chimpanzee. *Science,* 1980, *208,* 313–314.

Wittgenstein, L. *Philosophical investigations* (G. E. M. Anscombe, trans.). New York: Macmillian, 1963.

CHAPTER 2

Apes and Language: The Search for Communicative Competence

H. Lyn Miles

In the years since Gardner and Gardner (1969), Patterson (1978), Premack (1972), and Rumbaugh, Gill, and Glaserfeld (1973) first demonstrated that apes could acquire a set of symbols (gestural signs, computer lexigrams, or plastic tokens representing words or ideas), controversy has arisen over the extent to which the apes' use of these symbols could be called language. Objections to the belief that apes could possess language have ranged from a priori assumptions of animal inferiority, stemming from the western worldview that only humans possess a rational soul, to specific problems with the methods, evidence, or conclusions of the ape language researchers (Brown, 1973; Limber, 1977; Mounin, 1976; Terrace, Petitto, Sanders, & Bever, 1979). Review and criticism of primate language research from linguists, psychologists, anthropologists, and philosophers is a necessary and important means of furthering our understanding of the abilities of apes and the nature of language and intelligence. However, some criticism has done more to obscure this quest than to illuminate it.

For example, critics such as Chomsky have taken a viewpoint that examines only part of the evidence. Chomsky's nonevolutionary perspective denies a biological continuity between human and animal communication. He stated:

> When we ask what human language is, we find no striking similarity to animal communication systems . . . human language, it appears, is based on entirely different principles. This, I think, is an important point, often overlooked by those who approach human language as a natural, biological phenomenon. (Chomsky, 1972, p. 70)

Indeed, there are special features of human language that have not yet been identified in animal communication. That does not mean, however, that there is no biological relationship between the two. Unless one takes a creationist view, human language evolved from earlier hominid communication systems, which in turn share origins with ape communication in the hominoid systems of the Miocene. A comparison of the relative language abilities of apes and humans will help to determine the extent to which these may be homologous and may also help us understand human origins. It is no longer enough to simply assume differences between apes and humans. We must attempt to explore and explain the relative abilities of both.

There has also been misunderstanding about the nature of sign language and how it is reliably interpreted. Savage-Rumbaugh, Rumbaugh and Boysen (1980), who use the computer lexigram system with chimpanzees, have criticized projects that use sign language by pointing out the possible confusion of natural gestures made by apes and true manual signs. Some confusion can occur in evaluating the signs of both apes and human children. However, fluent signers or trained observers using reliability checks can successfully distinguish signs. Researchers have also devised means for conservative interpretation of sign language. For example, Hoffmeister (1977) has shown that all pointing gestures should be interpreted as the *point* sign until additional linguistic evidence suggests that a research subject specifically means *this* or *that, here* or *there*.

Other critics have argued that apes do not sign at all but are inadvertently (or deliberately) cued to perform certain movements by their trainers (Sebeok, 1980; Umiker-Sebeok & Sebeok, 1981). Charges of deliberate deception are serious and require careful substantiation. Unconscious nonverbal cues, or what is called *paralinguistic expression*, provide message redundancy and important behavioral and social information in both animal and human communication. It is a factor that must be taken into consideration when doing research, not only with apes, but also with human children—a point sometimes underemphasized by critics of primate language research. There is no question that paralinguistic expression occurs in the communication between signing apes and humans. As these critics point out, the double-blind testing of apes to date has not totally eliminated every conceivable source of unintentional cues.

There is evidence, however, that the signing behavior of apes cannot be exclusively attributed to paralinguistic cueing. First, the preliminary double-blind testing of apes using all three communication systems has shown that apes can label objects in the absence of obvious cues. Second, the cueing employed with Clever Hans, a horse at the turn of the century who appeared to count and answer questions, was a simple message for the horse to start and stop tapping a hoof. The sign constructions made by apes are too complex to be controlled by a simple on-off signal. Third, the orangutan in our laboratory will attempt to communicate with meaningful signs when alone with humans who do not know sign language and therefore could not possibly cue the correct signs. Finally, even Sebeok and Umiker-Sebeok (1979) concede that the total elimination of cueing during testing may be an impossible ideal. Two-year-old children with equivalent language skills might themselves fail such a restrictive test.

Perhaps the most frustrating critical approach to ape language experiments has been the assumption that language can be characterized as a single ability and that the critical task is to determine whether or not apes possess it. This approach treats language like virginity—you either have it or you don't. There are several limitations to this approach. First, it leads to a review of the ape projects with a shopping list of necessary and/or sufficient characteristics of language. The goal is often not to compare ape and human abilities, but rather to safeguard human uniqueness. Proponents of this approach usually begin by selecting one aspect of language, such as syntax, referential ability, conversational skills, etc., and proceed to show that apes lack it. This has resulted in an interesting game of linguistic "Yes, but . . . ." It begins with ape researchers who claim that their animals use signs spontaneously. Critics reply, "Yes, but does the animal form these into combinations—language after all, is not isolated calls or signals," and then, "Yes, but are these sequences governed by grammatical rules," and, "Yes, but do they use dependent clauses," and so on until some domain of language is found that is exclusively human.

The question of human uniqueness is an interesting one, but my point is that this approach to ape language research misses a great deal. It forces all concerned to guess at the essential element of language ability with a lot of competition over whose guess is best. It fosters the overinterpretation of some evidence by the media and some investigators. This puts undue pressure on ape language researchers to "prove" their animal has acquired whatever the latest linguistic fad determines is the essence of language. It has resulted in a sometimes bitter controversy among investigators who feel pressured to show that their method, species, or specific animal is more advanced than the others. Premack expressed this dilemma:

> Ten or so years ago when I first gave colloquia on the question, "Do chimpanzees have language?," we used to be booed and hissed. The colloquia had the character of a revival meeting. Then, after about five years, people were asking me, "Do your chimps do so and so?" and I'd say, "No," and they'd say, "Other people's do. You must be some sort of a dummy." (quoted by Bazar, 1980, p. 4)

Most ape language investigators agree that apes, and possibly some other animals, can learn a subset of skills related to human language and that these skills are not equivalent to adult human language. The difficulty lies in determining when a type of communication can be called "language." As Gardner and Gardner (1971, p. 181) pointed out: "There are ways to define language that would permit us to say that Washoe achieved language in April 1967. There are other ways to define language and undoubtedly the term could be defined in such a way that no chimpanzee could ever achieve language." The problem lies in the conception of language as a single "thing." Defining language is like trying to identify an elephant while holding only its tail or trunk.

The phenomenon we call language actually consists of many communicative and cognitive processes that require the coordination of several areas of the brain. It involves understanding that a word represents an object or idea, processing the sounds or visual representations of sounds into meaningful units, recognizing the rules of human discourse, structuring words to form sentences, and coordinating the articulatory apparatus to produce speech. Even if there are unique linguistic features that are present only in humans, these features operate interdependently with other processes. This is supported by evidence from studies of language acquisition that show that these interrelated abilities emerge within their own developmental sequences, eventually resulting in what we call adult language (Bates, 1979). In fact, Brown (1973), who has developed a widely accepted set of stages of language development, argued that children in the earliest stage of language development are not using adult language. At this stage, children base their constructions exclusively on meaning relationships, not on the grammatical rules with which they later learn to structure their sentences. This has also been found to be the case with apes, who by conservative measures do not appear to have moved beyond these earliest stages of language development (Bates, 1979; Brown, 1973; Miles, 1978).

Pursuing a specific level of skills necessary to claim language abilities for apes is a fundamentally less fruitful approach than the task of carefully describing and analyzing the development of cognitive and communicative abilities exhibited by these animals, regardless of whether or not it meets someone's definition of language. One way in which this can be accomplished is by taking an analytical and developmental approach. This approach abandons dichotomies in which apes are said to possess or not to possess language, and instead attempts to identify and analyze the various cognitive and communicative processes that underlie language (Bates, 1979; Miles, 1978). This approach has a strong basis in studies of language acquisition in children. Recently, investigators have become interested in the cognitive and social prerequisites to language and how nonlinguistic factors contribute to the development of linguistic structures. Some investigators have stressed the relationship between language and other aspects of symbolic functioning and the gradual development of language from nonlinguistic processes (Bruner, 1975; Piaget, 1962). Others have suggested that social interaction and the growth of communicative competence play a role in the emergence of language abilities (Bates, 1976, 1979; Bloom, 1973; MacNamara, 1972). Bates, through her studies of the emergence of communicative behavior in children 6 months to 1 year of age, has concluded that the superficially separate domains of language, thought, and communication are homologous. According to Bates (1979, p. 3), ". . . the symbolic capacity emerged phylogenetically from a combination of cognitive and communicative capacities that were preadapted in the service of other functions."

By looking in detail at the emergence of symbolic behavior in apes we can identify patterns of social and cognitive development that are associated with different degrees of success in acquiring language. This approach to ape language research has at least two advantages. First, a developmental approach to understanding protolinguistic processes may also help to identify the cognitive and communicative abilities that were significant in the evolution of language and a symbolic capacity. Those abilities that are within or just beyond the range of the great apes are possibly preadaptations for language development in our species; those that are beyond the reach of apes are likely to be ones that were significantly selected for in human evolution. Thus, a developmental approach in ape language research allows us to investigate the small continuous changes in communicative capacity that may have led to unique human abilities. Second, a developmental approach focuses attention on more specific and detailed studies of the language performance of apes. It allows us to ask many second-generation questions about ape abilities, such as, What does an ape "intend" when it uses a sign or computer lexigram? Are ape signs merely associative or true symbolic representations? Can apes use their signed communications to initiate individually conceived spontaneous messages? Hopefully, interest in these questions will end the "great ape debate" and allow investigators and observers to engage in more fruitful dialogue about the language capacity of primates.

Project Chantek

It is the goal of Project Chantek to begin to answer some of these new questions, which should lead to a fuller understanding of the processes that underlie language. My purpose is to investigate the communicative, cognitive, and semantic processes that underlie the emergence of symbolic behavior by analyzing the linguistic development of Chantek, a 3-year-old orangutan. All recent investigations of the language abilities of apes have employed the African apes, especially chimpanzees. Chimpanzees have been used most often because of beliefs concerning their superior abilities and the fact that they are more easily obtained for research than other apes. The orangutan is an endangered Asian ape now limited to the islands of Borneo and Sumatra. In contrast to the more social gorilla and chimpanzee of Africa, the orangutan leads a relatively solitary and arboreal life. For a long time, orangutans were considered to be less intelligent than the African apes and they were generally characterized as lethargic and unresponsive in captivity (Maple, 1980). However, in contrast to their reputation, orangutans generally score higher on cognitive tests than do gorillas or chimpanzees (Lethmat, 1978; Maple, 1980; Rumbaugh & Price, 1962; Yerkes, 1916). An orangutan is an interesting subject for our research because of this evidence of greater intelligence and also because the linguistic abilities of orangutans have not been systematically investigated.

Chantek (Malaysian for "beautiful") was born at the Yerkes Primate Research Center on December 17, 1977, and he arrived at our project in October 1978, when he was 9 months old. My goal in raising Chantek was to provide a comfortable environment, but not to attempt to rear him totally like a human child. I wanted to retain natural orangutan behaviors and provide at least a partially arboreal environment for him to develop as normally as possible, within the constraints of our research program. From the time of his arrival, Chantek has been housed in a five-room house trailer on the campus of the University of Tennessee, Chattanooga. The trailer is situated in a courtyard enclosed by chain fencing. The courtyard area includes a large jungle gym, picnic table, barrels, and several trees. Because orangutans are often arboreal, we strung numerous hanging ropes from tree to tree and to the building adjacent to our courtyard. Chantek's trailer is equipped with standard household items plus a variety of toys and playthings. Two walls of his bedroom were replaced with chain fencing. His bedroom contains a large wooden jungle gym and numerous ropes and hanging objects that have been tied across the ceiling to provide additional structures from which he can hang. Suspended from his jungle gym is a sleeping hammock, with various "nesting" materials such as towels and sheets, which allows him to sleep off of the ground. Chantek readily accepted weaning, some diapering (later he was toilet trained), and a diet prescribed by the Yerkes Center nursery. For the first 6 months Chantek received 24-hour/day care from myself and a small staff of caregivers. It was important to the goals of the project to keep Chantek's "family" small, but it was equally important that his caregivers interact with him in a relaxed, calm manner, since orangutan mothers are quite permissive in the wild. For the first 6 months we took turns sleeping with Chantek at night, much as his natural mother would. Then I slowly began to shape Chantek's behavior so he would allow us to move away from him at night and eventually leave his trailer after he had fallen asleep.

Chantek's sign language training was broken down into three stages. First, we established the rules for human communication and discourse; second, we taught him specific gestural signs which we brought to criterion for acquisition; and third, we sought to encourage his independent use of signs to transmit meaningful information about himself and his environment. In the past, some investigators have assumed that language skills are based solely on the construction and decoding of linguistic propositions. Recently, researchers have recognized the role of communication in language acquisition (Bruner, 1975; Greenfield & Smith, 1976). Part of the task of learning language also involves understanding and following a set of rules for noticing, sending, and receiving messages; that is, in language acquisition one must learn the rules or social conventions of language as well as how to structure the linguistic propositions that are expressed. Investigators have shown that the communicative abilities of human children seem to precede the ability to construct a linguistic proposition. In order to develop Chantek's communicative competence we familiarized him with some basic requirements of communication, such as paying attention and learning to take turns. We carried out several games and procedures designed to emphasize these skills. The next step was to introduce gestural signs to Chantek. For this we used pidgin Sign English, a form of sign language that uses the gestural signs of ASL in English word order without articles or extensive use of grammatical morphemes. We taught Chantek a new sign by molding his hands into the proper form and repeatedly associating it with the appropriate object, event, person, or concept (e.g., direction, color).

Motivation is a key factor in ape language research. Chantek was trained to use signs when he seemed most attentive or involved in the activity. We watched his behavior closely to determine his interests. We encouraged Chantek to use signs but did not pressure him or maintain a strict acquisition schedule. We stressed that his signs had behavioral consequences. He was not prodded into signing with repeated requests to name objects or other items in his environment. Instead, when he made a sign, caregivers attempted to respond appropriately or indicate that they could not comply. For example, if he obviously wanted a drink but signed *food-eat* we offered him something to eat or told him that no food was available. This reinforced his use of signs as meaningful communications and showed him that signs had social consequences and that caregivers would pay attention to his gestures. It was my belief that Chantek's communicative competence would be enhanced more by this method than by pressure from caregivers to use particular signs they required. When these animals are pressured they typically produce strings of signs until they accidentally hit upon the one that is correct in the caregiver's view. In this respect, Chantek's caregivers were companions, not trainers or instructors. I believed that if Chantek were to produce meaningful communications his caregivers would have to respond to him in meaningful ways. Chantek could not be viewed simply as a subject required to complete a series of trials in which his task was to find the correct sign. Repeatedly naming objects may give the appearance of language acquisition in that the animal's vocabulary increases, but it does not allow the animal to initiate the use of signs in spontaneous communication.

I was also interested in what factors would encourage signing in an orangutan. Would he be most interested in food, play, or in signing about his environment, or for that matter, would an orangutan sign at all if not constantly provoked by a "trainer"? I tried to interest Chantek in his natural environment and encourage communication about daily events, such as a cardinal at our bird feeder, a cat nesting under the trailer, or someone walking by with an ice cream cone, one of Chantek's favorite treats. During automobile rides, I encouraged him to direct me to different locations or indicate that he wanted to *go*. Later, I placed food items in various locations on campus so that he might "forage" for his food and ask to be taken to the *banana tree*, or tell me during a walk on campus that an apple was just up ahead. Sign language was a distinct advantage in encouraging this environmental reference since it is portable and did not limit us to communication within his trailer.

Initially, each occurrence of Chantek's signs was recorded, but, because of the frequency of his signing, beginning in November 1979 only the different signs and combinations and the context in which they occurred were recorded. In September 1979 we began to videotape a 1-hour monthly sample of relaxed sign exchanges between Chantek and a caregiver. These sessions occurred in our playroom laboratory, in Chantek's trailer, outside in his courtyard, or during walks on campus. Our goal was to provide a relatively natural and varied sample of the development of Chantek's signing.

Chantek produced his first signs *food-eat* and *drink* after 1 month of training, in December 1978. By June 1981 he had acquired 56 signs according to our acquisition criterion of spontaneous and appropriate usage on one-half of the days of a given month. These signs are listed in Table 2-1. The growth of Chantek's vocabulary through January 1980 is illustrated in Figure 2-1. As Chantek acquired new signs he continued to utilize his older vocabulary. In his second month of training, he began spontaneously to combine signs into sequence such

Table 2-1. Signs Meeting Acquisition Criterion (January 1979 to June 1981),
Listed in Order of Acquisition

food-eat	banana	flower	dog
drink	raisin	Lyn	cheese
come	key	listen	me
brush	bread	Kim	meat
tickle	hug	cereal	hat
go	Ann	ice-cream	cat
up	apple	berry	share
more	cracker	out	jacket
open	point	puppet	monkey
hurry	orange	John	bird
chase	touch	peach	bad
Chantek	milk	dirty	coke
nut	toothbrush	good	red
candy	give	Jackie	Richard

as *come food-eat* (in English glosses multiple meanings of signs are indicated by
hyphens). By his third year of training these combinations had grown in length
to sequences such as *puppet tickle point* (for me to tickle his bottom with his
puppet); *listen Ann go* (for a caregiver to carry him toward a loud bird); and *key
milk drink open* (to have his trailer door unlocked so he could go in and get a
drink).

It also became apparent that the manner in which Chantek executed his signs
was different from the signing chimpanzees I studied at the Institute for Primate
Studies in Oklahoma. Unlike chimpanzees, Chantek rarely made immediate repetitions of his signs. His overall manner of signing was also slower and more articulate. A comparison of videotaped conversations with Chantek and Ally, a signing chimpanzee, showed that Chantek signed less than one-half as fast as Ally
(Chantek, .92 seconds/sign; Ally, .39 seconds/sign). I believe this could represent
a species difference, since orangutans in their general behavior tend to be slower
and more deliberate (Maple, 1980).

Communicative Competence of Chantek and Nim

Recently, Terrace and his colleagues (1979; Petitto & Seidenberg, 1979; Seidenberg & Petitto, 1979) mounted an effort to show that apes cannot acquire language based on the results of a 4-year study with Nim Chimpsky, a chimpanzee
who received sign language training in their laboratory. They did not take a
developmental approach, but rather sought to show that their animal lacked
several essential elements of language. Although Nim learned to use a vocabulary
of 125 signs, Terrace concluded that in contrast to human children, Nim's multisign combinations were repetitions of less complex sequences that did not
increase in length nor in grammatical complexity as his training progressed. He

also concluded that Nim could not engage in human discourse. Terrace et al. (1979, p. 900) reported that Nim interrupted and imitated his teachers and, "...had not learned the give and take aspects of conversation that is evident in a child's early use of language." Furthermore, while children learn to initiate conversation by 3 years of age, 86% of Nim's communications were only in response to those of his "teachers" (Sanders & Terrace, 1979). Most children gradually decrease their tendency to imitate adult utterances, but, in contrast, Nim actually increased his rate of imitation. Terrace and his colleagues (1979)

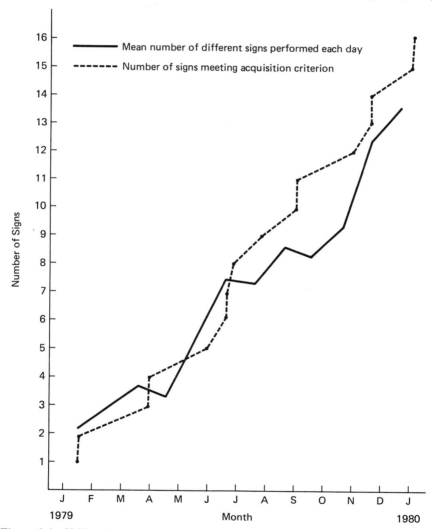

Figure 2-1. Utilization and growth of Chantek's vocabulary, January 1979 to January 1980

also argued that Nim was not the only ape to fail to show language abilities. They claimed that films and reports of several signing animals (Washoe, Koko, and Ally), ". . . showed a consistent tendency for the teacher to initiate signing and for the signing of the ape to mirror the immediately prior signing of the teacher" (p. 899). They argued that while apes can learn large vocabularies of signs or other symbols, they cannot produce original sentences nor can they master even the rudiments of conversational exchange within the ability of human 2-year-old. They concluded that, ". . . unless alternative explanations of an ape's combinations of signs are eliminated, in particular the habit of partially imitating teachers' recent utterances, there is no reason to regard an ape's multisign utterances as a sentence" (pp. 900-901).

Since one of the goals of Project Chantek was to investigate the relationship between language and communicative competence, I sought to compare the performance of Chantek and Nim. Such an analysis could provide additional information about ape communicative competence and also permit a comparison between the language capabilities of two species of apes. Terrace had based his conclusions on a sample of Nim's conversational exchange or discourse with one of his teachers when Nim was approximately 26 months old. We took a similar sample of Chantek's discourse at the same age, when he had acquired 26 signs. (Because Nim's training began a year earlier than Chantek's, Nim had acquired 40 signs at 26 months of age; however, at a comparable point in training Nim had acquired only 11 signs compared with Chantek's 26.)

Relatively relaxed interactions between Chantek and one of three caregivers in a 17 by 27 foot playroom were videotaped in February and March 1980 as part of our regular program of videotaped sampling. The playroom was furnished with a wooden jungle gym, toys, chairs, and a cabinet stocked with food and other items. Chantek was allowed to choose various activities, play by himself, or communicate with his caregiver. He was not pressured to sign or otherwise perform during these sessions. It was my goal to obtain as natural a sample of Chantek's ordinary signing as possible. For five ½-hour signing sessions, the caregiver signed and interacted normally with Chantek. This provided the discourse sample of Chantek's signs that was analyzed in the same manner as Nim's. For two additional sessions, the caregiver refrained from using sign language but responded appropriately to Chantek's use of signs and otherwise acted normally. Transcripts of all of the tapes were prepared and the reliability of the transcription was checked by having an assistant independently transcribe the first 10 minutes of each session. The average agreement between the two transcriptions was 91%.

Chantek's general level of language development was evaluated by analyzing the corpus of his sign communications obtained from the 2½ hours of mutual signing with a caregiver. Brown (1973) devised a measure of stages of language development based on the growth of the child's MLU and the size of the longest utterance. These measures were adapted for the acquisition of sign language by Hoffmeister, Moores, and Ellenberger (1975), and for convenience sign communications are sometimes referred to as "utterances." Studies have shown that a child's utterances become linguistically more complex as their mean and longest length gradually increase (Bloom, 1973; Bloom & Lahey, 1978; Brown, 1973). A child at the earliest level of language acquisition (Brown's Stage I) has an MLU less than 2.0 with a longest utterance of typically 4, but no more than 7 words (Brown, 1973). Terrace et al. (1979) reported that Nim's utterances in his discourse sample did not resemble those of a human child. At 26 months, Nim's MLU was 1.6, which in Brown's schema would place him in Stage I. However, unlike Stage I children, Terrace reported that Nim's MLU did not increase. Nim also continued a practice of repeating and stringing signs together in long sequences that did not resemble sentences. His longest utterance of 16 signs was far beyond the limits of a Stage I child. However, our analysis showed that Chantek's MLU was 1.93 (range 1.73-2.14) with a longest utterance of 6 signs. This places Chantek well within Stage I boundaries. Furthermore, as shown in Table 2-2, Chantek's MLU has steadily increased to the upper limits of Stage I since the beginning of the project. Unlike Nim, Chantek did not regularly string his signs together in repetitive sequences. Examples of his longest utterances from these sessions included *Ann tickle point* (to his foot), *give more food-eat apple*, and *tickle come tickle Chantek*. Chantek's multisign combinations have not yet been analyzed for any evidence of grammar or rule following behavior, so there is no evidence that these sequences are sentences. However, human children in the earliest stage of language acquisition do not produce sentences either; rather they combine related concepts into brief sequences similar to Chantek's.

In addition to combining words, children must also learn the give and take of conversation, including turn-taking and other skills of discourse. To determine Chantek's conversational competence, a discourse analysis adapted from Bloom

Table 2-2. Length of Chantek's Signed Utterances (January 1979 to May 1980)

Month	MLU	Longest utterance
January 1979	1.00	2
February	1.01	2
March	1.03	2
April	1.02	2
May	1.03	2
June	1.07	2
July	1.05	3
August	1.08	3
September	1.06	2
October	1.08	3
November	1.52	3
December	1.08	3
January 1980	1.51	4
February	1.68	9
March	1.79	7
April	1.74	7
May	1.91	5

and Lahey (1978) by Terrace and his colleagues (1979) was carried out. In this method, each of Chantek's sign utterances was categorized on the basis of the nature of his caregiver's prior utterance. Chantek's utterances were classified as one of the following: (1) an utterance that contained some or all of the signs of the caregiver's utterance (imitation); (2) an utterance that contained some of the signs of the caregiver's utterance with the addition of some new signs (expansion); (3) an utterance that contained none of the signs of the caregiver's utterance (novel); and (4) an utterance that was not immediately preceded by a prior caregiver's uuterance (spontaneous).

In our sample of mutual signing with a caregiver, Chantek produced 371 utter-

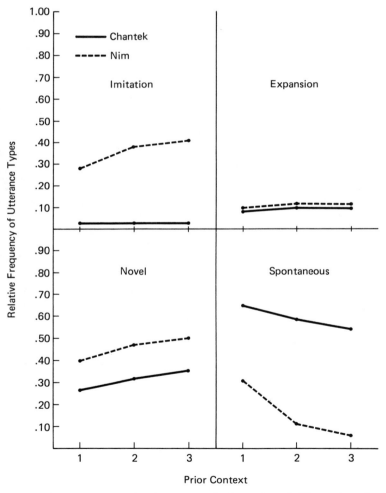

Figure 2-2. Comparison of the discourse of Chantek and Nim using the first, second, and final prior context

ances that could be classified according to their relationship to the prior caregiver utterance. An utterance for which the prior context could not be determined (for example, if the caregiver was out of view of the camera) was not included in our statistical analysis. The results of this analysis are compared with those reported for Nim (Terrace et al., 1979) in Figure 2-2. Terrace claimed that Nim would not converse but instead merely imitated his partner. By the most conservative prior context measure (the "back 3" or final context), Nim's rate of imitation was 38%, while only 3% of Chantek's utterances were imitations of his caregiver. Bloom and Lahey (1978) reported that the Stage I children in their study had an imitation rate of 18%, which decreased as the complexity of their utterances grew. Thus, Chantek's low rate of imitation is within the boundaries of Stage I language acquisition. However, Nelson (1980) argued that the imitation rates of children are more variable. Therefore, Terrace et al. (1979) may be premature in excluding Nim from Stage I language.

In addition to a low rate of imitation, our investigation also showed that Chantek was much more spontaneous in his signing than Nim. Only 8% of Nim's utterances were spontaneously initiated by him, whereas 37% of Chantek's utterances were spontaneous (Figure 2-2). Bloom and Lahey (1978) reported that the average proportion of a Stage I child's utterances that are spontaneous and unprompted is approximately 31% with a range of 22%-47%. This places Chantek squarely within the Stage I range.

The utterances of children also include expansions of adult utterances, often with the addition of new information. Bloom and Lahey (1978) reported that 10%-28% of the utterances of a child in the earliest stage of language acquisition are expansions of the adult's prior utterance. The rate of expansion for Chantek and Nim was somewhat low (6% and 8%, respectively); however, there are important differences in Chantek and Nim's expansions. Terrace et al. (1979) reported that Nim's expansions contained only a small number of nonspecific signs that did not add new information. In contrast, over one-half of Chantek's expansions added new information by making a reference specific to the conversational context. For example, when his caregiver asked, *you want more?*, Chantek replied, *more food-eat cracker*. When his caregiver asked, *food-eat what? which (food)?*, Chantek answered, *give food-eat nut*. Chantek also used over one-third of his vocabulary in his expansions, while Terrace reported that Nim used only a few new signs to expand his teacher's utterances. This is significant because, as Savage-Rumbaugh et al. (1980) suggested, adding new information by making specific reference is a crucial ability in distinguishing an associative versus symbolic use of signs or lexigrams by apes.

Chantek's high rate of spontaneous signing prompted us to videotape the two additional sessions in which his caregiver purposely refrained from using signs with Chantek. This allowed us to determine the extent to which he would sign if he was unable to imitate the caregiver or be prompted by the caregiver's signs. Terrace's experience with Nim suggests that these conditions would quickly extinguish Chantek's use of signs since he could not imitate his caregiver. However, we found that during these two sessions Chantek produced 133 utterances, at the rate of one utterance every 27 seconds. These utterances included *more raisin bread, come chase Chantek,* and *go drink.*

Terrace et al. (1979, p. 897) also reported that Nim frequently interrupted his teacher and that such interruptions ". . . detract from the conversation since they result in discourse that is simultaneous rather than successive." Therefore, we also examined the extent to which Chantek interrupted in our laboratory. We found that during the conversations both Chantek and his caregivers produced approximately the same number of utterances; Chantek produced 391 and his caregivers produced 418. Thus, in terms of the give and take of discourse, the conversations were relatively balanced exchanges. We also found that Chantek interrupted approximately as often as he was himself interrupted by his caregiver. In the 2½ hours of interaction, 8% of Chantek's utterances were interruptions of his caregiver, and 9% of the caregivers' utterances were interruptions of Chantek. In contrast, Sanders and Terrace (1979) reported that 35% of Nim's utterances were interruptions. This is significantly higher than Chantek's rate $(\chi^2 \ (1) = 75.52, \ p < .01)$. Terrace et al. (1979) presented data showing that Nim's later interruption rate was as high as 51%. However, again, Terrace may be too critical of Nim's performance. Nim's failure to emulate the ideal of a well behaved middle-class child is hardly grounds to deny Stage I language. Interruptions by apes and children can also reflect social and cultural factors that affect interaction. Even within one linguistic community, parents can vary widely in the extent to which they demand polite turn-taking in conversations. Thus, the cause of Nim's high interruption rate may be the behavior of his teachers rather than his lack of linguistic potential.

Conclusions

There are at least three factors that may explain the differences between Chantek and Nim. First, they may be due to behavioral differences between orangutans and African apes, who have been separated by many millions of evolution. In prehistoric times, orangutans ranged throughout Asia and were more numerous than they are today. Orangutans are known for their tool-using ability in captivity and are reported to have on occasion dismantled their own cages (Maple, 1980). Chantek's slower, more deliberate, and spontaneous signing and the distinctive "insightful" cognitive style of orangutans may indicate superior linguistic and cognitive abilities. Relatively little is known about the behavior genetics of the great apes. However, comparative biochemical evidence suggests that orangutans are less closely related to humans than the African apes, and it would be puzzling if their abilities proved to be vastly superior to those of gorillas and chimpanzees. Further research concerning the comparative abilities of apes will be necessary to resolve this issue.

A second explanation rests with the differences in Chantek and Nim's training environment. Nim had a group of over 60 "teachers" in his "classroom," consisting of an 8 by 8 foot concrete room. Terrace gave these reasons for this training environment:

> The room used as Nim's classroom was bare and small, a mere eight feet square. This was by design. I felt that Nim would not romp around too much in a small area and would be more likely to concentrate on the activities introduced by his teachers. I also felt that a bare room would minimize distractions. (Terrace, 1979, p. 897)

Although Nim had a more stimulating environment at the large mansion in which he was housed, this classroom for language learning was an unfortunate choice for a young ape needing a varied and stimulating environment. It is also inconsistent with a large body of knowledge that shows how important stimulating environments with color and movement are to infant learning. Apes, like children, can withdraw in a boring environment. Terrace described the effect this had on Nim:

> The slightest noise caused him to hoot and leap into the arms of his teacher. At times Nim was so scared that he tried to hide under his teacher's skirt . . . Almost every day, the [laboratory] rats' high pitched squeals evoked momentary panic in Nim. He would . . . rock back and forth on the floor. When he stopped rocking, he would frequently have lost interest in what he was doing before he was startled. (Terrace, 1979, p. 50)

Terrace also failed to recognize the tremendous importance of establishing rapport with Nim, his "informant" in this language experiment. This may be due to the fact that almost none of the members of Project Nim had any previous experience in training or handling great apes. Terrace explained that:

> Nim's teachers proved themselves not by how well they played with Nim or by how well Nim liked them but by how well they taught him to sign. Since their time with Nim was short, they devoted most of their classroom sessions to advancing Nim's knowledge of sign language. (Terrace, 1979, p. 54)

In fairness to Terrace, it should be understood that staffing an ape language project can be extremely difficult because of the skills that are necessary, the "earthy" nature of the work, and the tremendous amount of time that must be spent with the animal. On the other hand, it is not possible to have large groups of people who spend relatively little time with an ape come in and immediately discipline the animal and teach complex communicative skills. In addition, Terrace admitted that his project was plagued with arguments about when and how Nim should be disciplined, how he should be taught, and who really understood and loved him the most. This type of conflict cannot help but have a disruptive effect on an animal exposed to so many different sets of expectations, teaching styles, and affectional needs.

The specific linguistic methods used to train Nim were also peculiar. At first, Nim was trained using a method developed for retarded individuals in which the subject engages in repetitions of various tasks broken down into small stages. These procedures were inappropriate for Nim's level of neurophysiological development. Although apes can be used as animal models in applied language and cognition research, apes are not equivalent to retarded human beings. During the period in which this method was used, Nim not only did not acquire any new signs, he stopped signing altogether. This method actually helped extinguish Nim's signing behavior. Later, Nim was taught many signs with another peculiar method. Signs were taught to Nim in a vacuum without the presence of a referent. Terrace (1979, p. 52) described this method: "If Nim was to be taught the sign *cat*, it would be easier for him to practice making the sign *cat* . . . without the disruptive presence of the animal itself."

Despite Terrace's stated goal of raising Nim like a human child, no human child is initially taught language without reference to the real world. It is not difficult to understand Nim's lack of motivation to sign spontaneously about irrelevant concepts such as a nonexistent cat. All this succeeds in doing is teaching meaningless hand movements, so it is not surprising if Nim's conversations lacked human character. Nonlinguistic training methods cannot help but lead to a confused and nonlinguistic chimpanzee. In fact, it is remarkable that Nim performed as well as he did under these circumstances.

In contrast, the emphasis of Project Chantek has always been on first establishing rapport with Chantek and on the motivation for spontaneous communication. Language training cannot progress without a close relationship. Chantek's small group of caregivers have been with him from 10 months to 3 years. Ann Southcombe, one of Chantek's caregivers, has had several years of experience training gorillas, and I have formerly trained signing chimpanzees. Each caregiver spends a minimum of 15 hours each week with Chantek and is trained in ape-handling techniques. Signing is introduced as a primary means of communication *within* the social relationship rather than presented as lessons in a classroom. Chantek's learning environment is not limited to a small classroom but includes his home, the university campus, and frequent trips throughout the city and countryside. Chantek is never taught signs without their referents present. For example, in contrast to Nim, Chantek was taught the sign *cat* when he was curious about my cat Anushka and the stray cats that often wander into his courtyard. His interest in these animals and the strange sounds they made provided a meaningful context into which the sign could be introduced.

A final factor that may have affected the different performances of Chantek and Nim is the specific conditions of the conversations that were videotaped for the discourse analysis. Terrace's use of edited films, made to illustrate the signing of other apes, for a discourse analysis suggests that he does not appreciate the effect various contexts and purposes can have on the interaction and linguistic output of the participants. For example, there is a tendency to attempt to induce speech or signing in a laboratory conversation by prodding the animal to name objects in the environment. This results in a very unnatural conversation and efforts to avoid this were made in our laboratory. Social dominance can also affect a conversation. In order to evaluate properly an ape's performance, the linguistic and behavioral input to the animal must be examined.

The failure of Terrace and his colleagues to teach Nim communicative and linguistic skills should be viewed as the failure of one training method, not of the capacity of apes. We have shown that Chantek initiates communication with multisign sequences and that he signs spontaneously with a low rate of imitation or interruption of his caregiver. These results support the evaluation of the linguistic abilities of apes as comparable to the performance of human children in the earliest stage of language acquisition. Efforts to deny or claim abilities for apes on the basis of adult-level language should be abandoned. This approach should be replaced with a developmental perspective in which the cognitive and communicative abilities exhibited by apes are carefully described and analyzed, and in which apes are used as animal models in attempts to understand the emergence of communicative and symbolic capacities in ontogeny and phylogeny.

Acknowledgments

This work was supported by National Science Foundation Grant BNS 8022260, National Institutes of Health Grant HD 1491801, University of Chattanooga Foundation Grants RO4107609 and RO4107620, and University of Tennessee at Chattanooga Grant EO4107601. I gratefully acknowledge the loan of Chantek by the Yerkes Regional Primate Research Center, supported by National Institutes of Health Grant 00165. I thank K. Bailey, M. Beatey, D. Gebarowski, J. Miller, R. Miller, A. Southcombe, and M. Wofford for their assistance with Project Chantek. I especially thank C. Yeager and M. Biderman for help in analyzing data and T. Maple, J. Sachs, and P. Lieberman for their advice during preparation of the project.

References

Bates, E. *Language and context: The acquisition of pragmatics.* New York: Academic Press, 1976.

Bates, E. *The emergence of symbols: Cognition and communication in infancy.* New York: Academic Press, 1979.

Bazar, J. Catching up with the ape language debate. *American Psychological Association Monitor*, 1980, *11*, 4-5, 47.

Bloom, L. *One word at a time: The use of single word utterances before syntax.* The Hague: Mouton, 1973.

Bloom, L., & Lahey, M. *Language development and language disorders.* New York: Wiley, 1978.

Brown, R. *A first language: The early stages.* Cambridge, Mass.: MIT Press, 1973.

Bruner, J. The ontogenesis of speech acts. *Journal of Child Language*, 1975, *2*, 1019.

Chomsky, N. *Language and mind* (Enlarged ed.). New York: Harcourt Brace Jovanovich, 1972.

Gardner, B. T., & Gardner, R. A. Two-way communication with an infant chimpanzee. In A. M. Schrier & F. Stollnitz (Eds.), *Behavior of nonhuman primates* (Vol. 4). New York: Academic Press, 1971.

Gardner, R. A., & Gardner, B. T. Teaching sign language to a chimpanzee. *Science*, 1969, *165*, 664-672.

Greenfield, P., & Smith, J. *The structure of communication in early language development*. New York: Academic Press, 1976.

Hoffmeister, R. J. *The influential point*. Paper presented at the National Symposium on Sign Language Research and Training, Chicago, 1977.

Hoffmeister, R. J., Moores, D. F., & Ellenberger, R. L. Some procedural guidelines for the study of the acquistion of sign language. *Sign Language Studies*, 1975, *7*, 121.

Lethmat, J. Versuche zum "Vorbegingten" Handell mit Einem Jungen Orangutan. *Primates*, 1978, *19*, 727-736.

Limber, J. Language in child and chimp? *American Psychologist*, 1977, *32*, 1-13.

MacNamara, J. The cognitive basis of language learning in infants. *Psychological Review*, 1972, *79*, 1-13.

Maple, T. *Orang-utan behavior*. New York: Van Nostrand Reinhold, 1980.

Miles, H. L. Language acquisition in apes and children. In F. C. C. Peng (Ed.), *Sign language acquisition in man and ape: New dimensions in comparative psycholinguistics*. Boulder, Colo.: Westview Press, 1978.

Mounin, G. Language, communication, chimpanzees. *Current Anthropology*, 1976, *17*, 1-7.

Nelson, K. The first words of child and chimpanzee. In *Apes in language research*. Symposium presented at Southeastern Psychological Association, Washington, D. C., March 21-24, 1980.

Patterson, F. G. The gestures of a gorilla: Language acquisition in another pongid. *Brain and Language*, 1978, *5*, 72-97.

Petitto, L. A., & Seidenberg, M. S. On the evidence for linguistic abilities in signing apes. *Brain and Language*, 1979, *8*, 162-183.

Piaget, J. *Play, dreams and imitation in childhood*. New York: Norton, 1962.

Premack, D. Language in chimpanzees. *Science*, 1972, *172*, 808-822.

Rumbaugh, D. M., Gill, T. V., & Glaserfeld, E. C. von. Reading and sentence completion by a chimpanzee *(Pan)*. *Science*, 1973, *182*, 731-733.

Rumbaugh, D. M., & Price, C. Learning set formation in young great apes. *Journal of Comparative Physiology and Psychology*, 1962, *55*, 866-868.

Sanders, R. J., & Terrace, H. S. *Conversations with a chimpanzee: Language-like performance without competence*. Paper presented at the annual meeting of the American Psychological Association, New York, 1979.

Savage-Rumbaugh, E. S. Rumbaugh, D. M., & Boysen, S. Do apes use language? *American Scientist*, 1980, *68*, 49-61.

Sebeok, T. A. Clever Hans and smart simians. *Language in primates: Implications for linguistics, anthropology, psychology, and philosophy*. Symposium presented at Sigma Chi-William P. Huffman Scholar-in-Residence Program, Miami University, Oxford, Ohio, April 3-5, 1980.

Sebeok, T. A., & Umiker-Sebeok, J. Performing animals: Secrets of the trade. *Psychology Today*, November 1979, *13*, 78-91.

Seidenberg, M. S., & Petitto, L. A. Signing behavior in apes: A critical review. *Cognition*, 1979, *7*, 177-215.

Terrace, H. S. *Nim*. New York: Knopf, 1979.

Terrace, H. S., Petitto, L. A., Sanders, R. J., & Bever, T. G. Can an ape create a sentence? *Science*, 1979, *206*, 891-902.

Umiker-Sebeok, J., & Sebeok, T. A. Clever Hans and smart simians. *Anthropos*, 1981, *76*, 89-165.

Yerkes, R. The mental life of monkeys and apes: A study of ideational behavior. *Behavior Monographs*, 1916, *3*, 1-145.

CHAPTER 3

Chimpanzee Language and Elephant Tails: A Theoretical Synthesis

Roger S. Fouts

A large part of the controversy surrounding chimpanzee language research stems directly from the theoretical and a priori assumptions about the nature of behavior an individual experimenter brings to the research. Hence, the findings of those experimenters who assume a passive organism in their procedures may reflect the mental capacities and biases of the experimenter as much as they reflect the mental and linguistic capacities of the chimpanzee. On the other hand, those experimenters who assume an active, social chimpanzee are in a position to find the mental and linguistic capacities that reflect the chimpanzee's own abilities.

Results from the different ape language research projects appear to be contradictory if the differences in individual research procedures are ignored. Some researchers, such as Terrace, Petitto, Sanders, and Bever (1979) and Savage-Rumbaugh, Rumbaugh, and Boysen (1980), attribute the failure of their research to produce the results *they* expected to the biology of the chimpanzee rather than considering other aspects of their projects, such as training procedures, methods of data collection and analysis, and teachers. It is surprising to find this approach in experimental psychology, because a basic assumption of the discipline is that environmental stimuli control behavior. This assumption is translated from the general environmental situation into the laboratory with the approach that different experimental procedures produce different behaviors. What Terrace and Savage-Rumbaugh et al. have done is comparable to a Skinnerian who, wishing to produce scalloping behavior in the bar pressing of a rat, uses a variable ratio schedule of reinforcement, instead of the fixed-interval schedule of reinforcement that normally produces a scalloping effect, and then blames the rat's failure to show any scalloping on the mental capacities of the rat rather than looking to the procedures for possible biases (Savage-Rumbaugh et al., 1980; Terrace et al., 1979).

Terrace's goal was to train Nim to produce a sentence (Terrace, 1979; Terrace et al., 1979). He used highly structured drilling procedures in his training sessions to accomplish his goal. In these sessions the structural aspects of language were emphasized rather than the social aspects of language. As a consequence of Terrace's procedures, Nim's sentences were often mere imitations of the human trainer rather than truly spontaneous productions. The social elements of linguistic behavior that Terrace ignored in his training and data collection procedures were the very things he used to argue that Nim did not have language (the turn-taking that occurs in social conversations and the fact that Nim interrupted the human). Rather than addressing his procedural confounds, Terrace blamed the failure of Nim to meet his expectations on the chimpanzee's biology.

Recently, Yeager, O'Sullivan, and Autry (1981) examined Nim's conversational abilities under two conditions. One condition was modeled after Terrace's training sessions and a second condition was a conversational interaction in which Nim was allowed to sign about what *he* wanted to talk about rather than what the experimenters wanted to drill him on. Briefly, they found that the results of their training session condition were strikingly similar to the results Terrace reported, but the conversational condition was quite different in that Nim's spontaneous and novel utterances were greatly increased, whereas his imitations and interruptions were minimal. This study confirms the position that Terrace's findings were confounded by his procedures.

These apparent failures are failures only if the experimenters' preconceived goals are taken into account; they are not failures if they are viewed as being a direct result of the procedural methodologies. Instead, they tell us a great deal about what is important and necessary for language acquisition, in the same sense that a deprivation experiment tells us what is important for the natural development of a behavior. If a behavior does not occur when an organism is deprived of certain stimuli, then those stimuli are obviously very important in the development of that behavior.

In addition to telling us what is important for language development by depriving the organism of specific behaviors, the failures also tell us something positive: that is, that chimpanzees have the mental capacity to produce the structural elements of language even when the social and nonverbal components of language are absent or radically changed from a typical social interaction to a tutorial drilling situation. In other words, the experimenters' training procedures ignored or left out important elements of languaging behavior by emphasizing the structural aspects of language at the expense of the social aspects. Nevertheless, the chimpanzees in these projects learned certain aspects of language, even in the deprived situations.

A large part of the confusion and controversy results from the misconception that some researchers have concerning the nature of language. For example, Terrace (1979) implicitly drew on Skinner's very structured method of training while aiming for the grammatical structure of language proposed by Chomsky (1965). As a result, his experimental findings reflect very well what these two biases would produce if we really did learn and produce language according to these theories.

At this point, an attempt will be made to transcend the apparent discrepancies in the various results by proposing a theoretical conception of language that includes the results in their proper relation to one another. (For a complete discussion of this topic see O'Sullivan, Fouts, Hannum, & Schnieder, 1982.) The basic assumptions underlying this theory are that (1) organisms actively seek information, (2) language is largely a social behavior, and (3) language is an expression of the organism's cognitive process.

Because the proposition has been made that language is a behavioral expression of cognition, it is appropriate that cognition be discussed first. The view of cognition presented here also assumes a continuity across organisms as espoused by the theory of evolution, the basis and justification for doing any cognitive or learning research with nonhuman animals.

Theory of Cognition

Cognition is viewed here as a continuum ranging from behaviors that primarily require sequential processing to those that primarily require simultaneous processing. Sequential processing deals in causality, specific descrete acts, and temporal order; simultaneous processing deals in noncausality, gross motor acts, and the simultaneous organization found in patterns. For example, in regard to causality, for A to cause B it must precede B. However, if A and B co-occur, they can be perceived to be related to one another as some larger pattern, but neither A nor B could be considered to cause the other because of their co-occurence.

These two characteristics of the cognitive process have been traditionally dealt with separately in psychology, and, in fact, in the recent history of experimental psychology the sequential characteristics have been studied almost to the exclusion of the simultaneous characteristics of cognition. The reason for this emphasis within experimental psychology is that the sequential characteristics are much easier to measure and quantify than the simultaneous characteristics of the process. The emphasis on the sequential characteristics of cognition has led some scientists to assume that it accounts for all cognition. These scientists are victims of the same bias that the proverbial blind person suffered when describing the nature of the elephant by examining only the tail.

Psychology has suffered by its myopic focus on the sequential aspects of learning. However, this scientifically debilitating condition is not unique to psychology. A good portion of linguistics suffers from an analogous condition by focusing on the highly sequential aspects of language (the grammatical order of complete sentences) and ignoring or denying the importance of the simultaneous characterstics of language (the nonverbal social components of the regulators and meaning-carrying nonverbal characteristics of language). Linguistics, like experimental psychology, opted for what was easy to measure and quantify. In both disciplines a very narrow and sometimes misleading conception of behavior results from the active omission of critical aspects of cognition, such as simultaneity.

Language, like any other behavior, illustrates this cognitive continuum since it is made up, to varying degrees, of sequential and simultaneous characteristics. Written language is an example of a form of language behavior that is heavily loaded with the sequential characteristics of cognition. This emphasis on sequentiality is necessitated by the fact that written language is removed from the social context in which language normally occurs. As a result, the grammatical aspects of language become very important in order to remove the ambiguities that would normally be resolved by the social context. Even written language, however, involves simultaneous characteristics in the form of punctuation marks, such as the exclamation point and the question mark, which represent intonational information. The global transformations are also a product of the simultaneous process. From this point of view, the sentence is nothing more than a translation of the social utterance into the noncontextual rules of written language. In contrast, a conversation with a friend relies a great deal more on the simultaneous characteristics of cognition. Mehrabian (1968) claimed that as much as 75% of the meaning in a two-person conversation is nonverbal. At the extreme simultaneous end of the continuum is nonverbal language behavior. As with written language, extreme nonverbal language behavior has sequential characteristics in its production. For example, temporal order is very important in such nonverbal behaviors as turn-taking in conversation.

To understand fully this conception of cognition, it is best to examine its phylogenetic development. At one time in our early hominid origins, we probably had a cognitive system comparable to that of the extant chimpanzee. We were quadripedal and communicated mainly with one another using a gestural-intonational medium blended into nonverbal language behavior. When our hominid ancestors became bipedal, the hands were freed from locomotion duties and became more involved in gestural language behavior (Hewes, 1973). The number of gestures increased and became more discrete, just as there are discrete gestures in the (sometimes bipedal) wild chimpanzees' language behavior (Plooij, 1978). With the increase in the number of discrete gestures there was an increase in the precision motor movements required to produce these gestures.

These new discrete precision motor movements facilitated other behaviors that require this type of cognitive processing, specifically, tool making and tool use. Eventually this increase in the discrete precision movements of the hands also became involved in tongue movement. As with the hands, the tongue changed from the more generalized, cerebrally bilateralized control of continuous movements toward the lateralized control required to produce vocal speech. In order to produce vocal speech, the tongue makes discrete precise movements; all it is doing is stopping at specific places around the mouth just as the hands and fingers are stopping at specific places around and on the body when they are involved in gestural sign language. Darwin (1872/1965) noted the connection between precision movements of the hands and the tongue in observing that when people produce extremely precise movements, such as those required to thread a needle, their tongues often make sympathetic movements.

Once vocal speech occurred in our hominid ancestors, the tendency toward the present-day cognitive emphasis on sequential processing in adult humans began. It makes sense that the discrete motor movements of speech would come under the control of one hemisphere because there is only one tongue. If speech were bilaterally controlled, inefficient speech production would result; hemispheric competition has been found to cause stuttering (Jones, 1966). The tongue's singular nature would bias the control of speech behaviors to one hemisphere, and in most cases the hemisphere that dominates the tongue movements also dominates or controls the discrete precision movements of the hands (left or right). On the other hand, ambidexterity would not have the debilitating effect on precise manual movements that bilateral control has on the tongue movements of vocal speech because we have as many hands as hemispheres.

This conception of cognition can be graphically represented as two hypothetical curves over the cognitive continuum. One curve represents the typical human infant and nonhuman cognition, which is normally distributed between the two processes of sequentiality and simultaneity, whereas the adult human curve is skewed toward the sequentially loaded side of the continuum. The skewed distribution in the adult human is caused by the discrete precision motor movements required by speech and the immersion of an organism in an environment heavily loaded with sequential information (e.g., vocal speech). The skew toward the sequential end of the cognitive continuum results in the increase of by-products that also use sequential processing such as technology, written language, and cultural transmission of specific behaviors and knowledge.

The most profound change occurs early in an infant's development, when the brain has its greatest plasticity. As the organism grows older, the various parts of the brain become more specialized to perform functions. If the discrete motor movements of the tongue influence handedness, then this skewness would be even more exaggerated (Figure 3-1). The extremes of the continuum do not represent purely sequential or purely simultaneous characteristics; rather, the distribution should be viewed as degrees of blending. The sequential end has a greater proportion of sequentiality to simultaneity, and vice versa for the simultaneous end.

Note the relative position of the two means for the separate curves (solid lines) in Figure 3-1. It is the discrepancy between these two means that has traditionally led many scientists to believe that human cognitive processing is different in kind from animal cognitive processing. This conclusion is possible only if the scientist assumes that the adult human cognitive distribution is normal around its mean (Figure 3-1, broken line) rather than skewed.

The distribution of cognitive processing of the organism should also be represented in the neurology of that organism. Recent research on the distribution of gray matter (involved in sequential cognitive processing) in relation to the distribution of white matter (involved in simultaneous cognitive processing) in the two cerebral hemispheres has shown that the left hemisphere in right-handed adult human males has more gray matter than white matter (Gur, Packer, Hungerbuhler, Reivich, Obrist, Amarnek, & Sackeim, 1980). The greatest amount of

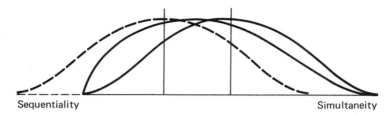

Sequentiality   Simultaneity

Figure 3-1. Cognitive continuum. The hypothetical normal curve with the mean on the right represents the typical human infant and nonhuman distribution of cognition which is normally distributed along the continuum between the two processes. The skewed curve with the mean on the left represents the typical adult human distribution of cognition which is skewed toward the sequential end of the continuum. The curve with the broken line represents a normal curve incorrectly based on the mean of the adult human curve which is actually skewed.

gray matter in the left hemisphere, as well as the greatest interhemispheric difference in the relative amount of gray matter, was found in the region that includes the basal ganglia, which are associated with fine sensory motor control. In other words, Gur et al. found that control of the discrete, precise motor movements involved in gestural language and speech is in the basal ganglia of the left hemisphere. The findings of Gur et al. lend strong neurological support to the proposed theory of cognition. These results also serve to explain away Chomsky's (1980) neurologically naive proposal that our species possesses a "language organ." The language organ Chomsky proposed is not an organ at all; it is simply the gray matter of the brain. Of course, language really requires both gray *and* white matter.

Assumptions About Cognition in Ape Language Studies

The proposed theory of cognition is useful in explaining the discrepancy in results from different chimpanzee language research projects. The various projects can be viewed as having different emphases along this cognitive continuum, with some stressing the sequential aspects of languaging behavior and others including more of the simultaneous aspects.

In the research by Gardner and Gardner (1971), the sequentiality of ASL interpenetrated the simultaneity of the socially enriched environment in which they raised Washoe. As a result their research was more of a blending, incorporating aspects of the entire cognitive continuum, than the sequentially biased approaches of the other researchers (Terrace, Rumbaugh, and Premack). Their method was to immerse Washoe in a rich social environment in which ASL took place in the context of daily events of her life in Reno: changing diapers, making beds, making and eating breakfast, playing in the sand box, going for car rides, playing hide-and-go-seek games and blind-person's bluff, receiving baths, and so on. Nearly all of the conversations (well over 90% as a conservative estimate) took place in spontaneous social interactions with her human companions (R. A. Gardner, 1980).

Perhaps the closest situation to a drill occurred in the comparison of methods of guidance used to teach Washoe some of her signs (Fouts, 1970, 1972). The structured teaching sessions in Fouts' study were conducted in two 12-week periods 1 year apart. The first 12 weeks were divided into four sections (two 4-week periods and two 2-week periods) separated by 3-week vacations. Each teaching session consisted of a 15-minute "settling in" period and 45 minutes of teaching. These 1-hour sessions occurred on Monday, Wednesday, and Friday of each week. The 12 weeks during the second year were grouped into three 4-week sections that were separated by 1-week vacations (for Washoe). The reason for the short training sessions and the vacations was that Washoe was a very young chimpanzee. Otherwise, they followed the same procedure.

The sessions took place in Washoe's 8 by 24 foot trailer. There were many things in the trailer that Terrace (1979) would consider distracting. The presence of distractions allowed Washoe a great deal more freedom than Terrace's procedure allowed Nim. She was allowed to take time out at will. She could always hide in a cupboard if the session got to be too much for her (as she did sometimes), or sit on her bed and play with her toys while ignoring the experimenter. She was also allowed to move around the trailer (often in the three-dimensional fashion of a brachiating chimpanzee), which meant some of the teaching was done on the run by the experimenter. The only other daily activity that might be misconceived of as drilling was the collection of reliability data. Even data collection, however, was interspersed with an ongoing activity: Washoe was asked to label an object during an activity that included that object. Signs that did not fit into an activity would usually be evaluated in a game, such as asking Washoe the name of a picture of an object, and some signs were not asked at all if they did not spontaneously occur in the normal course of Washoe's day, or if she did not feel like answering the question when asked out of its typical social context.

The above procedures are in extreme contrast to the method Terrace (1979, 1980) decided to use. Terrace (1979) stated that during part of his project Nim was taken to a classroom at Columbia University for 3-5 hours/day. The classroom could be considered a true classroom in only the most euphemistic sense of the word. It was a small (8 by 8 foot) room, painted white, that only had one window, but this was not a window to look through; it was a one-way mirror used to hide a videotape camera. Terrace felt that a variety of objects in the room would be distracting for Nim, and he wouldn't learn the signs on which he was being drilled. Washoe's trailer, in contrast, was well-stocked with toys and nooks and crannies that only a chimpanzee could fit into. The 3-5 hours/day of tutoring is again an extreme contrast to Washoe's 45-minute session, 3 days/week for 12 weeks/year for 2 years.

Gardner and Gardner tried to make things interesting and childlike with toys and games. They painted jungle scenes on the walls of the garage that Washoe had taken over as one of her favorite places to play. Toys, games, sand boxes, jungle gyms, tire swings, and interesting human companions (friends and surrogate parents and siblings) were as much a part of Washoe's life as they are in that of a normally reared human child, if not more so. Gardner and Gardner now have real playrooms with jungle scenes and toys and games for Moja, Dar, and Tatu to enjoy. Terrace claimed that Nim enjoyed a similar environment at the Delafield estate on the Hudson and when he was raised by Stephanie LeFarge. The exceptions were his excursions to Columbia for his 3-5 hour training sessions. If Terrace's suppositions were correct, that novel things in the environment are distracting and that highly structured drill sessions are necessary for language acquisition, then no human child would ever acquire language. Terrace's distracters are common in a human child's environment and are the essential enriching elements.

Terrace emphasized the structural aspects of language in the drill sessions at Columbia and in the structured videotaped sessions at Delafield. In these sessions, the emphasis was on getting Nim to produce sentences or signs, rather than merely conversing on a normal, social, and spontaneous conversational level. Because of this, Terrace's data (taken solely from these sessions in 3½ hours of videotape) reflect his own procedural bias toward the structure of written language and deemphasize the nonverbal and social interaction aspects of person-to-person language behavior.

The research done by Rumbaugh, Gill, and von Glaserfeld (1973) with Lana is another example of the sort of bias produced in the Terrace experiment. Lana was kept in a small plexiglass cage in which she had access to the keys she was required to push in specific sequences in order to receive her food and any entertainment that might be allowed by the experimenter. Unfortunately for Lana, she did not even have Delafield, as Nim did, as a respite from the drill sessions. However, she did have friends such as Gill who would go in and play with her. Once again the structural aspects of language were emphasized and the nonverbal and social aspects of language were deemphasized. (It is very difficult for a computer to smile or change the intonation of its voice or signs.) The social interaction at a spontaneous level occurred outside the data collection and training situations.

Premack's (1971) research can be viewed as very similar to that of Rumbaugh et al. (1973) in emphasis. Premack felt that Sarah had to be kept in a sturdy metal cage because she was 6 years old and therefore not tractable (according to Premack). Once again the social enrichment was absent or, at best, only occurred when someone took the time to form a social relationship with Sarah. Of course, the constituents of language that Premack considered important were the only aspects of language on which Sarah received her training.

These four research projects can be placed on a continuum from a sequential structural emphasis to a simultaneous or social language behavior emphasis. Premack's project would be at the extreme sequential end, followed closely by Rumbaugh's project, which would in turn be followed by Terrace's project. The project by Gardner and Gardner would be more to the middle or slightly toward the simultaneous side, with its emphasis on the social aspects of language behavior rather than the structural aspects.

These biases in procedure are clearly reflected in the results of the different projects. The results from the Terrace, Premack, and Rumbaugh projects demonstrated the sequential aspects of language but failed to demonstrate the social aspects. The results from Project Washoe demonstrated both the sequential and social aspects. When these projects are taken together, it can be seen that chimpanzees are within the range of language behavior of humans and therefore have the capacity for language.

Chimp Teaches Chimp

Finally, the research project on which I am presently gathering data suggests that human intervention may not be necessary for a chimpanzee to acquire signs. This study is examining the process by which a chimpanzee mother transmits her acquired sign language to her infant. It is an observational study, as opposed to the highly structured human-imposed training sessions found in some of the previous studies mentioned. The main control is that the humans use only seven signs in the presence of Washoe's adopted infant (*who, what, want, which, where, sign,* and *name*), and otherwise they use vocal English to communicate with the chimpanzees. The initial results of this research have been extensively reported by Fouts, Hirsch, and Fouts (1982).

The implications of the theoretical frame of reference presented in this chapter are that if language behavior or communication is to be understood, it should be studied by quantifying the development of a relationship from its beginning. The purpose of the ongoing research is to do this, with special emphasis being placed on the examination of its development in the mother-infant relationship. Because of unfortunate environmental situations, only the first 2 months of Washoe and her infant's relationship were examined before the infant died. The project has been able to continue after a 10-month-old infant, Loulis, was located for Washoe to adopt and raise in March, 1979. However, some very important months were missed because her adopted infant began imitating her signing 8 days after he was introduced to her.

In the initial analysis of some of the data, evidence of the teaching of signs by Washoe to Loulis has been found. On one occasion, Washoe was observed to place a chair in front of Loulis and then she demonstrated the chair-sit sign to him five times. However, we have never observed him to use this sign. Another sign, *food* (which he has in his repertoire now), was also actively taught by Washoe. On this occasion Washoe was observed to sign *food* repeatedly in an excited fashion when a human was getting her some food. Loulis was sitting next to her watching. Washoe stopped signing and took Loulis' hand in hers, molded it into the *food* sign configuration, and touched it to his mouth several times. Our initial impression was that beyond these two observations of behavior that appeared tutorial, most of Loulis' 17 observed signs were probably acquired through imitation of the other signing chimpanzees around him. However, in the past 2 months, as a result of a preliminary examination of some of the earlier data in more detail, evidence has been found indicating very subtle tutorial activity on Washoe's part. For example, when Loulis was first introduced to Washoe, Washoe would sign *come* to Loulis and then physically retrieve him. Later she would sign *come* and approach him but not retrieve him, and finally she would sign *come* while looking and orienting toward him without approaching him. Moerk's (1976) research suggests that human mothers may also actively teach language to their infants.

A behavior that appears to be an example of Loulis symbolically representing an object with an iconic gesture was videotaped. It occurred during a play session with a human. Loulis had a hard plastic bib and he would slide his bib under the cage toward a human who was unsuccessfully trying to grab it. This routine occurred over and over until the human successfully grabbed the bib from him. Loulis' response was to jump up and down and then touch his hands to his chest several times, and he was then given the bib. The data also show that at the onset Loulis was using his signs out of context (he was babbling). For example, when first placed with Washoe he imitated Washoe's signing *hat-George,* and then would jump, bob, and swagger and sign *hat-George.* Later, by the early part of 1980 he was starting to use signs in their correct context. For example, he now will point to an apple, signing *that*, and sign *gimme* over and over again. He has also been observed to use the *hat-George* sign to refer to all persons—chimpanzees or humans. The data for Loulis' signing is shown in Figure 3-2: the frequencies of signs, object manipulations, and manual babbling are reported for the months of July 1979, November 1979, and May 1980. Another description of the sign data

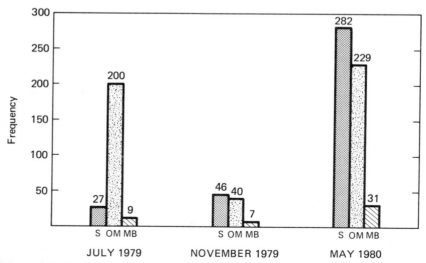

Figure 3-2. Frequency of Loulis' signs (S), object manipulation (OM), and manual babbling (MB)

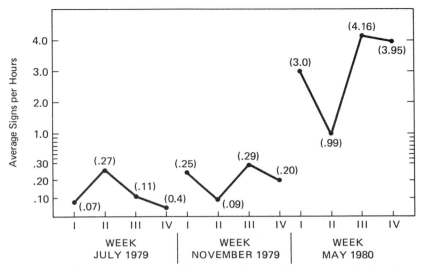

Figure 3-3. Average number of signs per hour of observation for Loulis

is given in Figure 3-3: the average number of signs per hour of observation are presented for the same months. In order to take into account the fact that the weekly hours of observation varied, the data represented in Figure 3-3 were computed by taking the total number of signs per week and dividing by the hours of observation for that week (Fouts, Hirsch, & Fouts, 1982).

Conclusions

The position I have presented is that language is primarily a social activity with a large nonverbal component. Language development, from this perspective, builds from preverbal (nonverbal) communication within significant relationships. Therefore, language acquisition in nonhuman primates must begin in infancy in the context of strong social bonds and utterances must be made in contexts meaningful to a young organism. Furthermore, the medium of communication should be compatible with the biology of the primate and, specifically, should take advantage of the predisposition of apes to communicate with gestures. The analysis of data must take into account the methodology of the experiment, and, in the case of ASL studies, the characteristics of sign language where it differs from oral language.

In addition to having a social basis, language use is a product of cognitive processes. Techniques of inculcation of language as opposed to immersion in a social milieu that includes language, and different linguistic systems, call into play different types of cognitive processes. I have tried to establish an association between simultaneous processing and nonverbal aspects of language use, and between the simultaneous nonverbal components and social development. In summary, I am proposing that the focus of some chimpanzee communication projects has been too much on syntax and not enough on creating the best environment for developing chimpanzee conversationalists; the data analysis has either overlooked this aspect of language use or has applied inappropriate measures out of ignorance of differences between oral and nonoral languages.

Generally, I am offering the concept of *continuity* as the solution to the problems of methodology and analysis in nonhuman primate language studies. There is continuity between language and other behaviors, both cognitive and social (e.g., tool making and nonverbal communication). There is continuity in the development of language, from so-called preverbal communicative behavior to adult language use. There is continuity from the evolutionary viewpoint, from protohominid social, cognitive and communicative behaviors to those of contemporary humanity. I reject the notion that there is some ultimate cut-and-dried criterion that distinguishes language from all other social and cognitive behaviors, or that distinguishes human communication and thought from that of all other species.

References

Chomsky, N. *Aspects of the theory of syntax*. Cambridge, Mass.: MIT Press, 1965.

Chomsky, N. Paper presented at the Apes and Language Conference, Houston, January 1980.

Darwin, C. *The expression of emotions in animals*. Chicago: University of Chicago Press, 1965, p. 235. (Originally published, 1872.)

Fouts, R. S. *The use of guidance in teaching sign language to a chimpanzee*. Unpublished doctoral dissertation, University of Nevada, 1970.

Fouts, R. S. Use of guidance in teaching sign language to a chimpanzee (*Pan troglodytes*). *Journal of Comparative Psychology*, 1972, *80*, 515-522.

Fouts, R. S., Hirsch, A., & Fouts, D. Cultural transmission of a human language in a chimpanzee mother/infant relationship. In H. E. Fitzgerald, J. A. Mullins, & P. Page (Eds.), *Psychobiological perspectives: Child nurturance series* (Vol. III). New York: Plenum Press, 1983, 159-193.

Gardner, B. T., & Gardner, R. A. Two-way communication with an infant chimpanzee. In A. M. Schrier & F. Stollnitz (Eds.), *Behavior of nonhuman primates* (Vol. 4). New York: Academic Press, 1971, pp. 117-184.

Gardner, R. A. Personal communication, 1980.

Gardner, R. A., & Gardner, B. T. Comparative psychology and language acquisition. In K. Salzinger and F. Denmark (Eds.), *Psychology: State of the art*. New York: New York Academy of Sciences, 1978, *309*, 37-76.

Gur, R. C., Packer, I. K., Hungerbuhler, J. P., Reivich, M., Obrist, W. D., Amarnek, W. S., & Sackeim, H. A. Differences in the distribution of gray and white matter in human cerebral hemispheres. *Science*, 1980, *207*, 1226-1228.

Hewes, G. Primate communication and the gestural origins of language. *Current Anthropology*, 1973, *14*, 5-12.

Jones, R. K. Observations on stammering after localized cerebral injury. *Journal of Neurology and Neurosurgical Psychiatry,* 1966, *29*, 192-195.

Mehrabian, A. Communication without words. *Psychology Today*, September 1968, 52-55.

Moerk, E. L. Processes of language teaching and training in the interactions of mother/child dyads. *Child Development*, 1976, *47*, 1064-1078.

O'Sullivan, C., Fouts, R., Hannum, M., & Schnieder, K. Chimpanzee conversations: Language, cognition and theory. In S. A. Kuczaj (Ed.), *Language development: Language, thought and culture* (Vol. 2). Hillsdale, N.J.: Erlbaum, 1982, 397-428.

Plooij, F. X. Some basic traits of language in wild chimpanzees. In A. J. Lock (Ed.), *Action, gesture and symbol: The emergence of language.* New York: Academic Press, 1978.

Premack, D. On the assessment of language competence and the chimpanzee. In A. M. Schrier & F. Stollnitz (Eds.), *Behavior of nonhuman primates* (Vol. 4). New York: Academic Press, 1971.

Rumbaugh, D. M., Gill, T. V., & Glaserfeld, E. C. von. Reading and sentence completion by a chimpanzee (*Pan*). *Science*, 1973, *182*, 731-733.

Savage-Rumbaugh, E. S., Rumbaugh, D. M., & Boysen, S. Do apes use language? *American Scientist*, 1980, *68*, 49-61.

Terrace, H. S. *Nim*. New York: Knopf, 1979.

Terrace, H. S. Paper presented at the Apes and Language Conference, Houston, January 1980.

Terrace, H. S., Petitto, L. A., Sanders, R. J., & Bever, T. G. Can an ape create a sentence? *Science*, 1979, *206*, 891-902.

Yeager, C., O'Sullivan, C., & Autry, D. *Communicative competence in* Pan troglodytes: *Rising (or falling) to the occasion.* Paper presented at the meeting of the American Psychological Association, Los Angeles, August 1981.

CHAPTER 4

Genes, Evolution and Language in Apes: The Nature of the Phenotypes

Edward C. Simmel

For several years now, and in a variety of graduate and undergraduate courses, I have had the opportunity to discuss with my students the linguistic adventures and accomplishments of Washoe, Bruno, Booee, Chantek, and the other talkative apes. The reactions of the students to this line of research have been quite varied, but generally seem to fit into four distinct patterns. A dreary minority merely write down whatever I say in vain hope of passing the next exam; a still smaller minority chuckle to themselves and mutter about circus stunts; a majority are enthusiastic, asking many excited questions about training procedures and the like; and one inquisitive student asks some searching and difficult questions about just where these investigations might lead. From this atypical student, one of the questions that is most intriguing to me can be stated as follows: If apes can so readily acquire a human type of communication—if they are indeed capable of language—why have they not developed this means of communicating among themselves, on their own, in their natural settings? Few answers, but a number of further questions and suggestions are generated by this question, which serves as the theme for the present discussion.

One partial answer, of course, is that all species of apes, in common with many other species, including humans, wolves, and mice, do communicate with each other in their natural settings. Such communication, by gesture, posture, and sound, is not restricted to alarms, threats, and requests by infants for food; rather, it conveys a large amount of fairly subtle information that includes mood and one's place in the social organization. Complex as such communication is, it is still quite removed from the more abstract form that we call "language." However, it seems reasonable that for most social species such relatively simple systems of growls, grunts, and gestures have worked well enough, and there has simply been little need (selection pressure) for anything more complex to develop. On the other hand, perhaps *our* early ancestors were so inept at sending or responding to simpler forms of communication that a more complex form had to develop for them to be able to survive to reproduce. Maybe this is all of the answer that we need: people talk as they do because they had to, in order to continue the species; apes didn't.

Such a glib answer would hardly satisfy most of us—certainly it would not (and did not) satisfy my inquisitive student. In the first place, we don't know, and can't know, what selection pressures led to the development of language in our species. We don't even know that there were such pressures. It is possible that language occurred as an accidental by-product of something else that was selected for. More important, we need to remember that evolution is not merely historical, but a continuing process. Therefore, the investigation of the evolutionary possibilities stemming from complex, abstract communication (language, if you will) in apes could serve as a splendid vehicle for viewing the role of complex behaviors in evolution, as well as aid the future investigation of a variety of ramifications of the communication process itself.

Natural Selection

In the modern synthetic theory of biological evolution, natural selection is no longer believed to be the only avenue by which all evolution takes place (Mayr, 1970). It is, however, a major factor, and the one of most direct relevance to our understanding of how environmental events influence systematic changes in biological and behavioral characteristics within species. In this brief review of the nature of the process of natural selection I will point out some common misconceptions without, I hope, creating any new ones. Following this, I return to the original topic and relate the essential ingredients of natural selection to what is known, and what must yet be learned about the uses of language in our primate relatives.

Two Misconceptions

In the simplest terms, natural selection occurs when certain members of a population are at a relative reproductive advantage over other members of that population, resulting in the modification of gene frequencies across generations. Thus, it is not, as the early evolutionists and many current lay people believe, *survival* of individuals that really counts, but rather reproductive success. This throws into question the many intriguing scenarios that some psychologists, ethologists, and popular writers have come up with to explain how some trait or other has "evolved" by providing its possessors with less chance of being eaten by predators, or swept out to sea by hurricanes. Such picturesque scenarios often sound quite believable, and it is tempting to use them. In doing so, however, one must assume, usually without any evidence, that those individuals not possessing the trait in question were regularly and heavily decimated by predators, or by hurricanes, prior to reaching reproductive age. While it is true that one must survive to early adulthood in order to reproduce, reproduction involves much more—fertility, mate selection (and acceptance), and the conditions needed for the viability of offspring, at the very least.

Another popular misconception about evolution, nearly the obverse of the notion of "mere survival" discussed above, can be termed the "use it or lose it" approach. The scenario here is quite simple, although totally wrong: Members of a species at one time possessed a trait that had no opportunity to be expressed. Being unused for a long period of time caused the trait to "evolve out"—it simply disappeared. This cannot happen. A characteristic that is truly not expressed (for example, the neurological potential for language in a species that has never had an opportunity to speak) can be involved in neither reproductive advantage nor disadvantage. The unused characteristic will, like the human appendix, continue in future generations of the species at the same rate as in the past.

Essential Ingredients

Natural selection operates on *phenotypes,* the more proper term for what we have been referring to as "traits" or "characteristics" thus far. A phenotype is simply the expression within or by an organism of some behavioral, physiological, or morphological characteristic. For example, size of canine teeth is a morphological phenotype; sympathetic nervous system activity preceding threat responses is a physiological phenotype; the baring of the canine teeth to a sudden intruder (part of a threat gesture) is a behavioral phenotype.

In order for natural selection to operate on any phenotype, it is essential that there be *variability* in the expression of the phenotype. In order for a phenotype to provide a relative reproductive advantage, it must be expressed to different degrees in members of a given population, or it may be present in some members and absent in others. The sudden onset of an environmental event that prevents reproduction within a population where all members are identical with respect to the key phenotype will result in the rapid extinction of that population. This unfortunate event would not occur if that phenotype were differentially expressed—only those members of the population not possessing the lethal phenotype would be able to reproduce. But what about the next generation?

This brings us to the third essential ingredient for natural selection: the *genetic basis* for phenotypes. There is little point in our belaboring the obvious: without genotypic variation underlying phenotypic variation, the "raw material" of natural selection, there could be no organic evolution. It might be worthwhile to consider briefly just what is meant and what is not meant by the term "genetic basis." This term simply means that one or more genes on one or more chromosomes are necessary (but not necessarily sufficient) for the development and expression of a phenotype; and, if there are differences in this phenotype within a population, some portion of the difference is the result of genetic differences. This is a straightforward biological explanation of individual characteristics being passed on from parents to offspring. A problem arises, however, when "genetic basis" is misused to imply that a phenotype is fixed, immutable, and innate. Although some phenotypes, blood groups for example, are set at the earliest stages of development and are rarely alterable thereafter, many other phenotypes (equally "genetic") are expressed differentially depending on the developmental stage of the organism, internal or external environmental events, or some interaction between these. When we know the genetic history of an organism and much about the environment in which it lives, many phenotypes are predictable; rather fewer are inevitable.

Which Phenotype?

Occasionally confusion arises regarding the distinction between *genetic effects* and *genetic differences*. There are many phenotypes that are invariant within any population but are clearly the result of gene action. Such phenotypes must be contrasted with those that vary within the population due at least in part to gene differences between individuals. In natural selection, the latter case is much more important than the former, for the reasons discussed previously.

Sometimes problems arising from the distinction between phenotypic similarity and phenotypic diversity (or, between genetic effects vs. genetic differences) result from ambiguity in the definition and measurement of the phenotype in question. For example, all humans have five fingers on each hand. This is obviously the result of the action of genes—a genetic *effect*—but is not a phenotype involved in evolution since there are no genetic (or phenotypic) *differences*. There are, however, a number of finger-related phenotypes on which there is considerable variation due largely to genetic differences, such as the size of the fingers, fingerprints, etc. These are potential candidates for involvement in natural selection if any of them provide a differential reproductive advantage. Of course, there are some characteristics of fingers, such as the number of callouses, on which individuals differ that are only minimally influenced by genetic factors and must therefore play little or no role in evolution.

To bring us back a little closer to our original topic, let us substitute the phenotype "language" for "fingers" in the preceding paragraph. All members of our species, with the exception of a trivial number who have gross abnormalities of one kind or another, are capable of spoken language. Clearly the product of the action of many genes, this is our invariant phenotype. There are also a number of language-related phenotypes having in all likelihood some degree of genetic involvement on which there are important individual differences among even those persons in highly similar environments: comprehension, total vocabulary, perhaps type of vocabulary, responsiveness to different types of words or phrases, etc. There are also aspects of language for which the differences can be attributed almost entirely to the environment, such as dialect or accent.

Phenotypes for Ape Language

By this time, my inquisitive student is becoming not only impatient, but pessimistic. Let us assume that apes are capable of some sort of humanlike language, that we have some pretty good guesses about how evolution takes place, and that putting these together would be likely to tell us something about apes, or communication, or evolution, or all three. The trouble is that there are so few apes available for study. They don't breed well in captivity, and even if they did, it takes so long for them to reach reproductive age. Since genetics is an essential ingredient in the evolutionary process, doesn't this bring our project to an unhappy end before it has even begun?

It is true that small numbers, long generation times, and breeding difficulties combine to make unfeasible the usual animal methods of genetic analyses such as inbreeding and selection. Similarly, pedigree analyses beyond more than one generation would extend beyond the patience of most investigators, to say nothing of the tenure committees at their institutions. Twins in apes are too rare to be of much use. Thus we have a rather gloomy picture of any immediate prospects for obtaining good quantitative data needed to establish genetic models— the number and location of the genes underlying the phenotypes we are interested in.

The history of science would be brief indeed if problems not yielding readily available solutions were quickly abandoned. There are often detours around "impossible" questions, and different ways of asking these questions. It is likely that this is the case here. Since quantitative genetic data are hard to come by, we must take a look at two other ingredients of the evolutionary process: variability and the nature of the phenotypes themselves.

Individual Differences in Ape Language

In the pioneering era of research on the development of sign language in chimpanzees, the focus of the work was properly on establishing the phenomenon and determining the most effective training techniques and achievement criteria. As these become established, it becomes possible to determine the extent of differences between individuals within species on various language-related phenotypes.

Individual differences can be measured meaningfully only for those subjects that are of approximately the same age, have relatively similar rearing histories, and have been trained and tested by identical methods. Differences within species are the primary concern for our purposes, although species differences are of interest, as is the difference in individual differences across species. A number of different aspects of ape language (language-related phenotypes) may vary differentially and may play different roles in life or reproductive success. The last point will be elaborated on at greater length later.

Although most of the work on individual differences remains to be done, in the mid-1970s Fouts and his students found suggestive evidence for individual differences in initial language (sign) acquisition in young chimpanzees (Fouts & Church, 1976). The number of chimps involved was small, and there were differences in their histories, but they were similar in that they had all been home reared and they were trained and tested in similar fashion. The results are promising enough to make this a path worth following, especially if a variety of phenotypes are measured.

The Nature of the Phenotypes

The most crucial questions stemming from all of our previous discussion come down to these: What *are* the phenotypes? How shall they be measured? In what ways do they relate to one another?

Before we can assess individual differences (phenotypic variability), it is obviously necessary to ask, Individual differences in *what*? Any determination of genetic models, difficult at best, would be absolutely impossible without a clear delineation and precise measurement of appropriate and unambiguous phenotypes. In other words, before anything can be known about the evolutionary significance of language in apes, we have to be very clear about just what it is that natural selection is acting upon.

Even apart from the genetic-evolutionary aspects, a "behavior-genetic" influenced approach—that is, an emphasis upon phenotypes—could enhance our understanding of the nature and functions of sign language in primate species other than our own. A few suggestions follow concerning a broadening of the sorts of phenotypes that could be investigated.

Thus far, emphasis has been on the *acquisition* of signs. Knowledge about the number of signs (total vocabulary) learned by an ape and the rate of acquisition is certainly essential. In addition, multivariate analyses of different measures of acquisition in the same subjects could give some indication of whether (or to what extent) various measures of acquisition constitute the same phenotype.

The major interest in, and controversy about, the development of language in apes has centered around cognition—the "intelligence" of apes, both in absolute terms and in comparison with humans. There are other categories of behavior, however, that might prove to be of greater significance for the functions and evolutionary significance of language. What role might emotional, motivational, or social factors play in the acquisition and use of language?

In order to increase our understanding of the potential importance of noncognitive factors in ape language, it will be necessary to develop phenotypic measures that go beyond mere acquisition. For example, there could be greater and more systematic determination of the types of "subject matter" most readily acquired (some individuals might be more interested in food, others in interpersonal relationships). Another type of measure that could get at the same type of problem might be some measure of spontaneity of signing: Do some individuals sign to humans or to other apes with different degrees of human encouragement, and if so, do they differ in what they sign about? Still another category of phenotypes relates to reactivity to signs: Are there some individuals who "speak" relatively little, yet are highly responsive to the signs of others? If so, what others? Under what conditions?

Broadening the scope of language-related phenotypes to include some that are other than cognitive in nature could add an interesting new dimension to the study of the development and use of language in apes. Our understanding of this new dimension would be further strengthened if these phenotypes and the relationships among them were, in turn, related to additional nonlanguage phenotypes that possibly tap similar categories of behavior (cognitive, emotional, social, etc.). Some of these additional phenotypes might be candidates for inclusion: emotional reactivity, both behavioral and physiological; curiosity and stimulus reactivity; perceptual and motor coordination; social reactivity; social rank (dominance); parental behavior; etc.

The relationship of these nonlanguage measures to how well apes "talk," about what and to whom, might yield nothing more than a random admixture of variables (also known by the technical term "mish-mash"); but, just maybe, some strong and consistent relationships would emerge. Should this be the case, we will have some predictors for certain types of language behavior, and, beyond that, the beginning of an understanding of the relationship of language in apes to noncognitive as well as cognitive variables. If the relationships among certain variables were strong enough, behavior geneticists might be willing to advance some tentative hypotheses regarding linkages, pleitropy, and common loci.

Looking further into the future, if sufficient longitudinal data were gathered from a sufficiently large number of apes, would we see any effects on reproductive success—mating, child-bearing, or infant care? Would we find that a particular type of language behavior affects mate selection? If so, and if it happened enough times, the result would be a change in gene frequencies—you know, evolution.

Acknowledgments

I am indebted to Margie Bagwell and Maria Lavooy for their editorial comments, and to Lyn Miles for her suggestions.

References

Fouts, R. S., & Church, J. B. Cultural evolution of learned language in chimpanzees. In M. E. Hahn and E. C. Simmel (Eds.), *Communicative behavior and evolution*. New York: Academic Press, 1976.

Mayr, E. *Populations, species, and evolution*. Cambridge, Mass.: Harvard University Press, 1970.

CHAPTER 5
Communication in Primates

Sarah Stebbins

For over a decade, experimenters have claimed some success in teaching chimpanzees to converse with human trainers and with each other in strings of plastic chips, electronically projected geometrical shapes, and signs from ASL.[1] Compared to most performing animals, the chimpanzees have mastered an impressive number of symbols, but none has anything approaching the massive vocabulary of a human language speaker. Utterances are typically short with little, if any, syntactic complexity. At best, researchers agree, the apes have exhibited the linguistic competence of very young children; but most anticipate more substantial achievements with additional time and the development of better training techniques.

Ape language research is commonly thought to have implications for the hypothesis that human language is species specific, learned through the operations of an inborn acquisition device. Moreover, experimenters claim that their work sheds light on other issues in comparative psychology and on the intelligence and behavior of nonhuman primates. Some critics, however, question whether the experiments show even rudimentary symbolic communication by nonhumans. The chimpanzees may be responding to nonverbal signals from their trainers, or they may be exhibiting conditioned responses to the presence of an object or in the expectation of a desired reward. Evidence for the spontaneous use of symbols in novel combinations or in novel circumstances is anecdotal, the critics complain, and the creative symbol combinations that are reported may be interpreted too sympathetically by hopeful trainers. Some researchers question whether all experimental subjects use symbols with the intention to communicate, and they design exercises to demonstrate that their own pupils do understand the nature and function of the symbols they use.

---

[1] More recently, Patterson (1978) claimed to have taught a gorilla to communicate in ASL. See also Miles, Chapter 2, this volume.

The debate raises a number of issues, but the first question for experimenters and critics alike is whether the chimpanzees are communicating with symbols in the laboratory. Experiments are designed against the background assumptions that correct responses are acts of communication of certain sorts. The most suggestive evidence from day-to-day interactions between subjects and trainers is found in reports of putative conversational moves: giving descriptions, making requests, asking questions, and even issuing insults. If the apes are not communicating, but are manipulating symbols in the course of some other sort of problem-solving activity, even the most modest claims must be suspended.

Parochialism in the Study of Animal Communication

I shall argue that both researchers and critics have parochial views of the requirements for communication. They misidentify the behavioral features relevant to communication, and this results in poorly designed experiments, counterproductive controls, and unwarranted criticism. Primate communicative behavior is mistakenly thought to require the semantic flexibility and self-awareness characteristic of human language use. In addition, it is mistakenly thought to be independent of the resemblance of laboratory conditions and symbolic channels to the contexts and forms of natural communication. I shall discuss the relevance of these features under the following three headings: natural contexts and channels, semantic flexibility, and self-awareness.

Before describing these errors and the confusions they generate, I would like to suggest a hypothesis about the origin of parochialism in the study of primate communication. I suspect that it is ultimately due to the tenacious influence of the Cartesian prejudice that nonhuman animals are machines, and that conscious human activity must be different in kind from any activity found in nature. Speaking a language is a paradigm of a conscious human activity, perhaps even what enables us to engage in any conscious activity; therefore, speaking a language cannot be different only in degree from the signaling found in insects, birds, and other social mammals.

Few would continue to endorse the view that animals are machines, but the assumption that linguistic communication is discontinuous with other forms of communication plays a role in delimiting the subject matter of ape language research. "Communication" is clearly equivocal; humans, higher mammals, and the most primitive organisms are all said to communicate. Communication is also ascribed to inanimate features of the natural environment, computers, and even rooms. From this heterogeneous collection, researchers must locate a class of related events that will bear on the understanding of human behavior. However, the behavior or features of many of these organisms and objects shed little light on the fundamental nature of other events and states counted as communicative according to some standard. Therefore, in practice, many researchers and critics adopt a more restrictive concept of communication to identify events relevant to their investigations. Although they explicitly question whether nonhumans can be taught language, they continue to regard linguistic communication as a singularity in nature. They consider speaking a language and behavior sharing a significant number of features with it to be described as "communicative" in one sense, and they consider all other events, including human nonverbal communication and complex forms of social behavior in other mammals, to be described as "communicative" in a different sense.

Perhaps, for some scholarly endeavors, the most productive distinction would be between speaking a natural human language and engaging in other sorts of behavior or exhibiting other sorts of states. I am convinced, however, that both comparative psychology and the study of nonhuman primate intelligence and behavior, as they are pursued in ape language research, would be better served by a broader concept of communication. What is needed is an understanding of communication as a particular sort of social behavior, evolved to promote group cohesiveness, locate willing mates, and facilitate cooperation and the nonviolent resolution of differences. The proper characterization of an interspecific concept of communication is notoriously problematic, and it is beyond the range of this discussion.[2] However, I do want to suggest that such a concept must have both a functional and an empirical, biological component. Communication can be recognized, in part, by the role it plays in the life of an organism and in the interactions it has with others. In addition, it must proceed, to some extent, from structures and processes with a certain evolutionary history. The interplay of the empirical and functional components will define a class of biological realizations of certain evolutionary strategies for maximizing benefits and minimizing risks of social interaction.

If these suggestions are correct and an appropriate concept of communication depends, even in small part, on biological structures and the behavior they give rise to in the field, there may be no simple answer to the question of whether the apes are communicating with trainers in experiments and in day-to-day encounters. Artifically created organisms and compounds often fall neatly into no natural category, and behavior induced in the laboratory may too fall on a borderline. It may be both significantly like communication and significantly different from it. This is not to say, however, that any behavior that takes place solely in the laboratory cannot be communication. Behavior that draws on and extends natural abilities and propensities to communicate, and that plays a similar role in the life of the organism as that played by natural forms of communication in the field, may well be considered communication.

These observations about communication, brief as they are, suggest some sorts of evidence to expect if the experimental animals are communicating in their use of symbol systems. One rather persuasive sign that they are communicating would be their attempts to use natural forms of communication in experimental situations when denied use of the symbol system, or when use of the symbol system is inefficacious. This would suggest that communicating is part of the chimpanzees' behavioral repertoire for solving the problems experimenters set, and that the behavior satisfies the functional criterion.

Another source of evidence would be the physical settings in which symbolic interchanges take place and the activities and channels the symbol system

---

[2] Wilson (1975) raised some difficulties with characterizing such a concept. See also Klopher and Hatch (1968) and MacKay (1972).

exploits. In general, the closer these conditions and features are to those found in natural forms of communication, the less skeptical we need to be. Of course, these judgments must be made against considerations of the general intelligence and adaptability of the experimental subject. Animal trainers know that the more intractable the creature, the less familiar and comfortable the circumstances, and the more unusual the behavior, the less likely it is that training will be successful. Few ape language researchers acknowledge this truism, and no researcher or critic recognizes the necessary link between these features and communication in ape language studies.

Natural Contexts and Channels

The concern to establish symbol systems and experimental contexts that exploit natural talents and propensities goes well beyond a concern with the brute physical limitations of subjects. Experimental results from the 1950s show that the vocal apparatus of chimpanzees is inadequate for the production of human speech and current research agrees to exploit the manual dexterity of chimpanzees in symbolic communication instead (Hayes & Hayes, 1951). Gardner and Gardner have gone further in adopting a particular symbol system, ASL, because of its resemblance to natural forms of communication in chimpanzees. They explained,

> More to the point [than manual dexterity], even caged, laboratory chimpanzees develop begging and similar gestures spontaneously . . . In our choice of sign language we were influenced more by the behavioral evidence that this medium was appropriate to the species than by anatomical evidence of structural similarity between the hands of chimpanzees and of men. (Gardner & Gardner, 1969, p. 664)

Manual dexterity enables chimpanzees to manipulate plastic chips and to use an electronic keyboard. The talents exploited here, however, are not as frequently reported in acts of communication by untrained chimps, and they are correspondingly less likely to be successfully directed to that purpose. A fortiori, there is good reason to doubt that pigeons communicate by pecking lights and buttons. Epstein, Lanza, and Skinner (1980) claimed to have trained two pigeons to request and give information about the color of a light visible to only one of them. Transmission takes place by the pigeons pecking keys labeled *What color?*, *R, Y,* or *G* (red, yellow, or green) and *Thank you.* Pecking is a frequent natural activity in pigeons, and proclivities to peck have been helpful in training the birds to perform a number of laboratory tasks. However, unless pecking occurs as a form of communication in untrained birds, there is good reason to suspect that the pigeons' exchanges only mimic communication.

Gardner and Gardner (1969) not only showed a concern with natural forms of communication, but they also attempted to establish laboratory conditions in which behavior that "could be described as conversation" would be developed. Unfortunately, they took as their paradigm of conversation typical interactions between young children and adults. They demanded that, in addition to requesting objects and attentions, Washoe ask and answer questions about objects. Insofar as possible, they wanted their chimpanzees to duplicate the linguistic performances of children in order to avoid the necessity of giving a behavioral definition of language. "If children can be said to have acquired language on the basis of their performance," they claimed, "then chimpanzees can be said to have acquired language to the extent that their performance matches that of children." (Gardner & Gardner, 1975, p. 244). However, the observations about communication above show this reasoning to be fallacious. Untrained chimpanzees have been observed to use gestures and vocalizations to request food, objects, and affection from conspecifics and from humans. In the field, they do convey information about their environment indirectly through warnings and gestures of reassurance. I know of no field observations, however, of behavior akin to asking questions and naming objects by apes. Communication does not seem to play that kind of epistemological role in their social life. Of course, they might exhibit this behavior in response to the frequent presence of human beings initiating these exchanges. In general, however, requests for food, tickling, and the like are better candidates for genuine communicative acts. (The frequency of these relative to other sorts of utterances by trained chimpanzees is evidence that the chimpanzees are communicating, rather than evidence that they are not.)

In field observations, participants in communicative exchanges usually have visual contact with each other, and are frequently close, sometimes touching. Ironically, such contact between trainers and subjects facilitates trainer cuing and courts the Clever Hans phenomenon. Clever Hans, the performing horse, tapped his hoof in answer to spoken questions; but, as the investigator Pfungst discovered, Hans responded not to the spoken words but to nonverbal signals from his questioners (see discussions by Sebeok; 1978a, 1978b). Direct sensory contact between subject and trainer allows the subject to detect subtle movements, changes in facial expression, breathing patterns, and the like, and to interpret these as guides to correct responses. A number of critics, most notably Sebeok and Umiker-Sebeok, have suggested that "talking" apes and dolphins may exhibit the Clever Hans phenomenon. "We believe," Sebeok and Umiker-Sebeok (1979, p. 91) stated, "that when a careful analysis is permitted by the researchers, much of the human-ape communication will be shown to be a product of inadvertent cues."

The Clever Hans phenomenon, Sebeok (1978b) suggested, consists of looking in the destination for what should have been sought in the source. Attention is misdirected in two ways. The channel of communication from trainer to subject is not what it appears to be; Hans responded not to the spoken question but to the the nonverbal signals his questioners offered. In addition, the subject's response is not what it appears to be; Hans appeared to be answering a question when, in fact, he was responding to a set of nonverbal commands by noncommunicative acts. Hans simply manifested "go and no-go responses to the minimal cues provided by the people around him," as Sebeok and Umiker-Sebeok (1979, p. 79) noted.

Patterson (1980) correctly noted that nonverbal cuing is to some extent involved in all human conversations. Facial expressions and bodily movements help to clarify spoken messages, and they are among the many sources of redundancy in human linguistic communication. They are sometimes independently sufficient for understanding, and no doubt there are occasions on which, without them, understanding would not take place even between fully competent and cooperative speakers. Nevertheless, this observation should not relieve the serious concern Sebeok and Umiker-Sebeok expressed about the Clever Hans phenomenon in ape language research. Their contention is that the linguistic content of questions matters very little, if at all, in provoking the correct response from subjects, nor is there any redundancy in spoken and cued messages in their view: trainers make two entirely different requests of subjects, and subjects respond only to the cued request, without performing an act of communication.

Ape language experimenters have attempted to guarantee that the identity of particular signs and symbols plays a larger role in the chimpanzees' symbol use by constructing "double-blind" experiments, in which the interlocutor does not know the correct response. These experiments require that the subject signal a computer, an observer untrained in the symbolic language, an observer sitting some distance away, or an observer watching through a one-way mirror. Sebeok and Umiker-Sebeok contended, however, that none of these measures completely rules out the possibility of nonverbal cues. In a particularly clever experiment designed by Savage-Rumbaugh, Rumbaugh, and Boysen (1978), two chimpanzees, Austin and Sherman, requested tools from each other to retrieve food from otherwise inaccessible hiding places. In order that one chimpanzee remain ignorant of where the food was hidden, they were separated by a plastic partition and communicated through electronically projected images. The authors reported that human supervision was frequently required, and Sebeok and Umiker-Sebeok noted that this is a source of possible cuing.

Sebeok and Umiker-Sebeok did not remark on the extreme artificiality of double-blind experiments as contexts for communication. The most persuasive evidence that Austin and Sherman used the symbol board to communicate comes from an occasion on which both chimpanzees were in a single enclosure. On this occasion they intermixed symbol use with touching, pointing, and extending an open hand, all natural communicative gestures (Savage-Rumbaugh et al., 1978, Author's Response, p. 615). Similarly, many of the less structured contacts between Washoe and her trainers offer better evidence of communication than the double-blind experiments.

Concern with the Clever Hans phenomenon has led to misinterpretation of a disturbing discovery by Terrace and his colleagues about the performances of signing apes. Terrace, Petitto, Sanders, and Bever (1979) claimed that in many ape-trainer conversations, the ape's use of a sign follows immediately on the trainer's use of the same sign. In a number of other cases, a sign is preceded by the trainer's use of a wh-sign: who, what, etc. This discovery offers very good reason to reexamine claims of vocabulary size and fluency, but it would be a mistake to represent this observation as directly supporting Sebeok and Umiker-Sebeok's claim that ape language performances exhibit the Clever Hans phenomenon. An introduction (*Psychology Today,* 1979) to an article by Terrace did just this, however: "Without the ability to create sentences independent of intentional and unintentional signals," it stated, "these primates cannot be said to be 'talking'." (Sebeok and Umiker-Sebeok, 1979, concur with this interpretation of Terrace's conclusions.) However, not all signals evoke a noncommunicative response by a subject.

We might make a distinction between cuing a response and prompting it. A cue is a command that may be issued through a channel distinct from that in which the apparent conversation takes place. An animal may, and most frequently in performance situations does, respond to a cue without communicating. A prompting word or gesture, however, must take place in the channel of the conversation; it is a candidate, or part of a candidate, communicative response. Thus, in imitating or otherwise responding as prompted, the subject does communicate. Adults frequently prompt young children in their conversations with them, and trainers may prompt acts of communication by chimpanzees. If Terrace has observed prompting, it may show an incomplete mastery by experimental subjects of symbol systems or the interplay of conversation, but it does not count against their having communicated in these contexts or in contexts in which they are not prompted.

Terrace himself regards even these problematic sign sequences as acts of communication. Within their limited capacities, apes use sign sequences to converse with their trainers. "Nim's comprehension of signs," Terrace (1979, p.252) reported, "made it possible to engage him in conversation with his teachers about many topics." It is somewhat misleading, therefore, for M. Gardner (1980, p. 3) to attribute to him the view that "there is no reason to suppose that [an ape] is doing anything essentially different from a pigeon that has been taught to obtain food by pecking four differently colored buttons in a specific order regardless of how the buttons are arranged." The bird, presumably, would not be communicating by pecking. Terrace et al. invited this misunderstanding, however, in their discussion of their suspicions that the chimpanzees' multisign utterances are learned serial responses. They wondered "what is to be gained by assigning names to each member of the sequence" (Terrace et al., 1979, p. 900). Moreover, in an early article on serial learning, Straub, Seidenberg, Bever, and Terrace (1979, p. 147) suggested that "in evaluating a chimpanzee's performance when it 'writes a sentence' of plastic chips or 'lexigrams,' it is important to consider whether those sequences are merely rote chains of responses in the service of an incentive."

Semantic Flexibility

Terrace's concern was not whether the chimpanzees are communicating, but whether they can produce sentences. Multisign utterances that are word-by-word responses to trainer prompting show no such ability. Even spontaneously produced strings are suspect; many can be interpreted as rotely learned sequences with a single variable. Terrace suggested that Premack's subject, Sarah, learned to substitute names for different foods into the symbol sequences *please machine give X* or *Mary give X Sarah.* The result of such substitution may fail to count as a sentence, in Terrace's view, for one of two reasons. The chimpanzees may be unable to make substitutions into other positions in the sequence, and they may not understand the meanings of all the words in the string. Terrace et al. (1979, p. 899) believed "it seems more prudent to regard the sequences of symbols glossed as *please, machine, Mary, Sarah,* and *give* as sequences of nonsense symbols."

An alternative, and more generous account, would treat the constant sequences as single, logically simple linguistic units, focusing on the symbol system as the chimpanzees use it rather than on the system Premack devised. For example, *please machine give apple* could be interpreted as a two-word string rather than a one-word request, *apple* preceded by semantically vacuous material. Terrace made the parochial assumption of a symbol-word isomorphism between the laboratory language and English, thus forcing the semantic structure of English onto analyses of the apes' utterances. Other researchers are equally guilty of this assumption, as is evident in their somewhat perverse translations of laboratory utterances into broken English. Even Gardner and Gardner, who, like Terrace, have adopted a human language in their research, translate the fluent signing of trainers into English strings lacking articles, prepositions, and the copula.

Obviously, a symbol system need not be isomorphic to English to be a language. Perhaps less obviously, a symbol system may lack some of the most fundamental semantical properties of human languages and still function as a tool for communication. For example, the significance of most items in human languages is independent of the nonlinguistic context in which the item is used, and most items are significant in every nonlinguistic context. However, some human signaling systems through which communication takes place lack this feature. Paul Revere's lanterns, for example, would not have meant anything placed in a different church steeple or on a different evening. More commonly, a double yellow line is only meaningful on a street or highway. Red and green lights mean one thing on a street, and something completely different on water or in the air. A nod may be a bid at an auction, a greeting in passing, or a signal of agreement in conversation. The context independence of symbol significance is not a necessary condition for communication.

Savage-Rumbaugh et al. (1978) mistakenly have taken it to be one, however. Their training procedures were designed to guarantee that tool names could be used in contexts other than the retrieving of hidden food, and that they were not directly linked with the act of using the tool to get the food: "If our tool names had been functioning at this level, the animals could not be expected to employ them to communicate with one another, because the lexigram words would only be functioning as performatives with their use limited to occasions where food was seen at the given tool site" (p. 542).

The ability to use a symbol in the absence of an object it represents is a special case of the nonlinguistic context independence of significance, and it is neither a necessary condition for a communication system nor a sufficient one. Thus Premack's (1972) features tests, in which Sarah demonstrates an ability to associate the properties of absent objects with their chip "names" are misconceived. They do not show that the relation between chip and object is exploited in communication. Sarah's performance can be described as solving a different sort of problem: making a transitive shift from the association of chips and objects to chips and other chips.

If the significance of words need not be independent of nonlinguistic context in a communicative system, then a fortiori the significance of words need not be independent of linguistic context. The free combinatory behavior of symbols, in which the meaning of a multisign sequence is a function of the fairly stable meanings of its parts, is characteristic of only some units in human languages, and it is not a necessary condition of the communicative use of a symbol system, nor is it a sufficient condition. Arithmetic functions and other mathematical algorithms share this feature, even when they are used as tools in calculation and not for the expression of results. A hand in a card game is somewhat like a sentence in this respect, as well; the significance of a hand is a function of the fairly constant significance of the individual cards and the structural relations they bear to one another. Moreover, play can share much of the give and take of human conversations. In contract bridge, particular cards may even be played for the sole purpose of communicating the content of one partner's hand to the other, but most card playing is not communicating.

Self-awareness

Researchers and critics sometimes put questions about the semantic relations between symbols and objects in terms of the mental states of the chimpanzees when using the symbols, but Savage-Rumbaugh et al. (1978) have been particularly adamant in insisting on a mental component to communication. They contended that their work demonstrates that "two chimpanzees have been able to comprehend the symbolic and communicative function of the symbols they use" (p. 539). They argued that other experimenters have not demonstrated such awareness, however. They complained, for example, that the limited range of chips available to Sarah for each task makes it unlikely that chips serve a symbolic function for her or that she is aware of their serving such a function. "[It is] difficult to understand how Sarah could come to realize that the plastic chips could be used to communicate desires and to control and orient the behavior of others," they objected. If Sarah was not aware of the function of the chips, however, she was "simply solving a set of problems" (p. 552).

Savage-Rumbaugh et al. (1978) also suggested that Gardner and Gardner, for other reasons, have not demonstrated Washoe's understanding of the function of her signing. They argued that the acquisition criteria Gardner and Gardner employed are so weak that, given the relatively iconic nature of ASL, "it is impossible to tell whether the chimpanzee is simply imitating or echoing, in a performative sense, the action or object, or whether the animal is indeed attempting to relay a symbolic message" (p. 551). They echoed Steklis and Harnad in suggesting that the transmission of a symbolic message requires that gestures be intended and relied upon to transmit information about referents. Savage-Rumbaugh's (1980) complaint about the Epstein et al. experiment was the same: "It's clear to me," she stated, "that Pigeon B does not know what color Pigeon A saw, or even that Pigeon A was looking at a color. The pigeons don't explain anything to each other. They don't know they're transmitting information. They're just pecking at colored lights."

It is no less parochial to require that all primate communication exhibit the inner, mental features of human linguistic communication than to require that it manifest the outer, observable features. Given what we know about bee dances, bees communicate the location of a food source whatever mental states accompany the insects' performances and observations. Communication between social mammals, and especially the primates, is well documented in field studies, yet we know very little about their mental life.

It may well be quite similar to ours. Griffin (1976; Chapter 11, this volume) suggested that complex mental states should be attributed to social mammals and insects on the grounds of parsimony. It is the simplest available explanation for their elaborate patterns of behavior, he contended. Premack and Woodruff (1978) argued that chimpanzees not only experience complex mental states, but attribute them to others as well. The versatility that chimpanzees show in communicating with humans and with each other using gestures appropriate to, and physical props found only in the laboratory suggests that conscious mental states may well play a role in devising communicative acts. Domestic mammals frequently show an ability to respond appropriately to communicative overtures by members of other domestic species. Presumably, studies of new interactions of wild populations will reveal similar versatility. These facts make it likely that some uptake mechanisms in social mammals are conscious, rather than nonconscious responses to genetic programing.

A popular philosophical tradition, founded on the work of Grice (1957, 1969), supports a conceptual requirement that communication involve self-awareness. Communication takes place, according to this tradition, when there is an act of meaning on the part of a transmitter and an act of understanding on the part of a receiver. An act of meaning, in turn, requires an intention to secure a belief, and possibly some further effect, by virtue of the receiver's recognition of the transmitter's intention. Human linguistic communication typically does involve these complex, reflexive intentions, as do many other forms of human communication; but this is better seen as a matter of empirical fact than as a conceptual requirement for communication.

Entertaining and recognizing reflexive intentions is an especially flexible mechanism for achieving successful uptake. Channels for communication may be consciously selected to optimize clear transmission, signals and codes may be varied at will to allow for very specific information transfer, and sophisticated judgments about the appropriateness of transmitting behavior can be made and acted on. Thus the possibility of unnecessary provocation and misunderstanding is reduced. There must be costs as well, however. A very complex mental apparatus is required, diverting energy from other sorts of development or increasing energy requirements. Responses are slower and less reliable, and attention is demanded that might be paid elsewhere. Other mechanisms for securing uptake would have other advantages and other costs.

Understanding human communication, in its full complexity, demands the recognition that many organisms have evolved related patterns of behavior that play similar roles in their social interactions. Communication, like courtship behavior, rearing young, and other natural activities, may be realized in different ways in different organisms, and different features may characterize the processes and products of these realizations. Human language use has long been thought to be especially significant in human mentality, but to take human mentality as a standard for the continuity of behavior with human communication is to exhibit a parochialism about communication that leads to serious misinterpretations of ape language research.

Acknowledgments

I am indebted to Marjorie Grene, Frederic Schick, and Robert Schwartz for their helpful comments on an earlier draft. I am also grateful to the Editors for their suggestions in preparing this chapter.

References

Chomsky, N. Recent contributions to the theory of innate ideas: Summary of oral presentation. In H. Morick (Ed.), *Challenges to empiricism.* Indianapolis: Hackett, 1980, pp. 220-240.

Epstein, R., Lanza, R. P., & Skinner, B. F. Symbolic communication between two pigeons. *Science,* 1980, *207,* 543-545.

Gardner, B. T., & Gardner, R. A. Evidence of sentence constituents in the early utterances of child and chimpanzee. *Journal of Experimental Psychology: General,* 1975, *204,* 244-267.

Gardner, M. Monkey business. *The New York Review of Books,* March 20, 1980, *27,* 3-6.

Gardner, R. A., & Gardner, B. T. Teaching sign language to a chimpanzee. *Science,* 1969, *165,* 664-672.

Grice, H. P. Meaning. *The Philosophical Review,* 1957, *66,* 377-397.

Grice, H. P. Utter's meanings and intentions. *The Philosophical Review,* 1969, *78,* 147-177.

Griffin, D. R. *The question of animal awareness.* New York: Rockefeller University Press, 1976.

Hayes, K., & Hayes, C. The intellectual development of a home-raised chimpanzee. *Proceedings of the American Philosophical Society,* 1951, *95,* 105-109.

Klopher, P., & Hatch, J. Experimental considerations. In T. A. Sebeok, (Ed.), Animal Communication. Bloomington, Ind.: Indiana University Press, 1968.

MacKay, D. Formal analysis of communicative processes. In R. Hinde (Ed.), *Non-verbal communication.* Cambridge, England: Cambridge University Press, 1972, pp. 3-25.

Patterson, F. G. The gestures of a gorilla: Language acquisition in another pongid. *Brain and Language,* 1978, *5,* 72-97.

Patterson, F. Letter. *The New York Review of Books,* October 9, 1980, 45-46.

Premack, D. Teaching language to an ape. *Scientific American,* 1972, *207,* 92-99.

Premack, D., & Woodruff, G. Does the chimpanzee have a theory of mind? *The Behavioral and Brain Sciences,* 1978, *4,* 515-526.

The trouble with ape-language studies, *Psychology Today,* November 1979, *13,* 63.

Savage-Rumbaugh, E. S. Quoted in *The New York Times,* February 19, 1980, section C, p. 2.

Savage-Rumbaugh, E. S., Rumbaugh, D. M., & Boysen, S. Linguistically mediated tool use and exchange by chimpanzees. *The Behavioral and Brain Sciences,* 1978, *4,* 539-554.

Sebeok, T. A. Ecumenicalism in semiotics. In T. A. Sebeok, *The sign and its masters.* Austin: University of Texas Press, 1978, 61-83. (a)

Sebeok, T. A. Looking in the destination for what should have been sought in the source. In T. A. Sebeok, *The sign and its masters.* Austin: University of Texas Press, 1978. pp. 85-106. (b)

Sebeok, T. A., & Umiker-Sebeok, J. Performing animals: Secrets of the trade. *Psychology Today,* November 1979, *13,* 78-91.

Straub, R. L., Seidenberg, M. S., Bever, T. G., & Terrace, H. S. Serial learning in the pigeon. *Journal of the Experimental Analysis of Behavior.* 1979, *32,* 137-148.

Terrace, H. S. *Nim.* New York: Washington Square Press, 1979.

Terrace, H. S., Petitto, L. A., Sanders, R. J., & Bever, T. G. Can an ape create a sentence? *Science,* 1979, *206,* 891-902.

Wilson, E. O. *Sociobiology: The new synthesis.* Cambridge, Mass.: Harvard University Press, 1975, Chap. 8.

CHAPTER 6
Limits of Primate Talk

Thomas W. Simon

Imagine the following conversation:

P: *What did you do to your friend?*
K: *Bite.*
P: *You admit it?*
K: *Sorry bite, scratch. Wrong bite.*
P: *Why bite?*
K: *Because mad.*
P: *Why mad?*
K: *Don't know.*
P: *I am sorry to hear that you are mad. Are you also afraid?*
K: *Yes.*
P: *Of what?*
K: *Don't know.*
P: *Are you afraid of me?*
K: *No, afraid, friend.*
P: *What else comes to your mind when you think fear? Does that image have anything to do with your friend?*

Who are the participants in this conversation? A child therapist and one of those menacing hyperactive children that seem to be overpopulating our schools these days? Or is this a cuckoo-nest dialogue between Ken Kesey and the Big Nurse? Or could it simply be an adult talking to a child—which some claim to be the simplest dialogic form imaginable? Would it be at all startling to discover that this is the first rendition of a linguistic interchange between a nonhuman primate and a digital computer? Although our example is contrived, it is not very contrived, consisting, as it does, of an amalgam of expressions from Weizenbaum's (1976) computer program ELIZA (see also Boden, 1977) and the gorilla Koko's recorded utterances (Patterson & Linden, 1981). By constructing this machine-ape communication have we bridged a gap greater than molecular biologists are doing between prokaryotes and eukaryotes with recombinant DNA?

Have we erected the ultimate Turing test (which seems to be a primary one for primatologists) whereby both parties can now be said to understand language because you, the observers, were fooled into thinking both of them were human?

In the past decade computers and primates have both become objects of linguistic investigation. In this chapter we examine the nonhuman primate side of these concerns. (Hereafter, "primate" will mean "nonhuman primate.") As will become repeatedly evident, however, we will not be able to leave the machines completely aside. Before temporarily abandoning the computers, please note one intuitively based judgment about the dialogue: The truncated patient-ape (K in the dialogue) responses look less like mature human adult linguistic responses than the complete sentence inquiries of the therapist-machine.

In any event, the present analysis of primate language studies takes us through two interrelated areas: methodological and ethical. The methodological and ethical issues are intertwined. Although the primate language research is defended against its most vocal opponents, other methodological problems with important ethical implications are pinpointed. Moreover, the consideration of various methodological alternatives such as human and computer studies raises serious questions regarding the continuance of this research.

Science of Primate Linguistics

Washoe's learning of some ASL initiated what now seems to be an endless definitional dispute over "language" (Gardner & Gardner, 1969; Umiker-Sebeok & Sebeok, 1980). The charges and rebuttals revolve around the necessary and/or sufficient conditions of language and whether nonhumans could possibly ever satisfy these conditions. Since the list is extensive, we concentrate on some of the more salient ones from the following areas: semantics, syntax, and physical structure.

Semantics

One would be hard-put to find critics who deny nonhuman primates the rudiments of semantics. Most agree that in order for a sign to be a symbol it needs to exhibit what Hockett and Altman (1968) labeled *displacement*, that is, to be "remote in time, space, or both from the site of the communicative transaction" (p. 63). In other words, the primate's use of a sign should not be tied to the immediate environment. Furthermore, use of the sign must be arbitrary. Washoe's signing of *banana* would seem to meet these minimal requirements—for an actual banana is not needed to elicit Washoe's sign.

However, the defectors from this position are found among the primate researchers themselves. Primatologists Savage-Rumbaugh, Rumbaugh, and Boysen (1978), who do not use ASL in their research, proclaimed that "there is no evidence, other than richly interpreted anecdote, to suggest that Washoe and other signing apes are producing anything more than short-circuited iconic sequences" (p. 551). Accordingly, successful use of ASL does not necessarily provide evidence of symbolic communicative activity since ASL has a high degree of iconicity and the animal might simply be recalling iconic gestures and imitating them.

What is disturbing about the above critique is not its truth in this particular instance but rather its truth in general. In other words, a deflationary redescription can be given of any of these primate studies. As a skeptic I could always reasonably reject your description of an event and substitute my own description of "what really happened." Even if presented with a film of Washoe signing with you I could then charge that off-camera cues were being used to coax Washoe to respond appropriately. Then after being allowed to view a filming session, I could still dismiss Washoe's feats as responses to subliminal cues from the experimenter. Through skeptical eyes Washoe quickly becomes a slightly more ingenious version of Clever Hans, whose equine mathematical skills were shown to be merely a matter of taking counting cues from the experimenter (Umiker-Sebeok & Sebeok, 1980). This scenario is endless, for in the long run there is no satisfying the dogmatic skeptic. Experimental control is never strict enough for the skeptic.

Reasons for skeptical success in these matters abound. Some of the more salient ones pertaining to the nature of the social sciences and experimental control will be addressed later. In the meantime we should note that part of the difficulty lies in the fact that these experiments involve ascribing internal states to the primates. These attributions are difficult to defend against the skeptic even when dealing with our own species, the difference being one of degree and not of kind. Thus, the skeptical critique applies to most experiments conducted in the social sciences and not just to the primate experiments. A behaviorist critique questioning the grounds for inferring internal states works for human and primate experiments.

It is critical to recognize that this dispute between skeptic and primatologist is not resolvable by experimental means (Gopnik, 1981a). More stringent experimental controls are not the answer. The disagreement reflects basic assumptions each side makes about humans and primates. The types of inflationary or deflationary descriptions used are a function of the nonsample information, including our value judgments. Subjective theory explicitly finds ways of including this nonsample information.

In fact, this particular sense of probability is called subjective because it allows us to include a priori judgments in probability assessments. A critical guideline to help determine the a priori assignments for animal experiments is the opposite of Lloyd-Morgan's oft-cited dictum that "the attempt to formulate a naturalist account of animal behavior should not be abandoned in favor of the easier road of anthropomorphism" (Pylyshyn, 1978, p. 592). The opposite is "Assume until proven otherwise that others are just as intelligent, complicated, and so on, in their own way as you are in yours. And be very skeptical of your own motives and intellect if you think you have proved otherwise" (Menzel & Johnson, 1978, p. 587). In other words, it is much more reasonable to be guilty of inflating animals' abilities than of deflating them. As Griffin (1976) noted, Frisch's lowly conception of the honeybee's capacities contributed to his delay in discovering the famous bee's dance for some 20 years. Therefore, while the dispute with the skeptics is nonempirical it is not also irrational. There are good reasons for rejecting the skeptic's charges. Objections can be raised against any experiment. Meeting these objections is often not a matter of stricter empirical control but rather a matter of rejecting certain philosophical assumptions in favor of others.

To show just how deflationary and protective an analysis can get at the semantic level, consider what Katz (1976, p. 37) called the "effability thesis" ("every proposition is the sense of some sentence in each natural language") which entails that all natural languages are intertranslatable. Since the languages of neither primates nor dolphins satisfy this condition, they cannot be said to be natural languages. Semantical studies of primate languages would therefore be doomed to failure. Overlooking controversial issues such as the indeterminancy of translation, this is nonetheless objectionable in its rigidity: whatever does not *now* qualify as natural language has little chance of ever qualifying. Dismissal by fiat is the ultimate skeptical position.

What is apparent from this discussion so far is that deciding whether primates have a semantics is no simple matter. Central to the dispute are difficult philosophical questions. However, we can at least draw one important conclusion. Semantical studies using primates cannot be summarily dismissed as unscientific.

Syntax

Even if we grant that the primates are using meaningful signs, syntactical problems still obtain. Brown (1970) asked, "whether Washoe was simply making signs distributed in time or whether the signs were in construction" (p. 224), that is, "sign combinations employed to express structural meanings" (p. 227). If Washoe and company cannot decipher the difference between *dog bite* and *bite dog*, then they are at a very early stage of childhood language development. However, the signing apes are able consistently to construct meaningful structures, for example, the subject–object order. Therefore, we need not worry about this particular level of development. Of course, we could always go on to claim, in ad hoc fashion, that this later stage of development still does not qualify as syntactic language even in the child, but that places us back on the skeptic's endless track.

A far more difficult syntactic complaint has been made:

> *All* of what Premack calls productivity in Sarah's use of language consists in her "generalizing" from the trained content . . . there is no indication that Sarah has ever done the most characteristic thing that a productive syntax permits human speakers to do; namely, use a sentence of syntactically novel form without being specifically trained on sentences of that form. Productivity in human language exploits iterative syntactic mechanisms which generate *novel constituent sequences*. (Fodor, Bever, & Garrett, 1974, p. 449)

One obvious reaction to this is to wait and see. However, just what constitutes "novel constructions" and whether humans actually employ these without any specific training are problems that need to be resolved before this is even taken as a challenge in the primate language studies. This is a rather typical problem, for the critiques often presuppose linguistic theories which are themselves questionable.

Before we can dismiss the primates' gestures as not having a syntactic structure we need to have a fairly clear theory of syntax, which, in turn, is embedded in a larger theory of language. Not only do we not have a generally accepted theory of syntax, but also, even if we did, how would we know that primates had a certain syntax structure according to that theory? Again, we are in the tenuous position of inferring internal states from external behaviors. The problems for primate language research in this regard are no different in kind from any form of human language research, particularly the child language acquisition studies.

In some cases, the proposed syntactic theory assumes that primates are incapable of certain syntactic feats. Chomsky (1966, p. 19) and his followers regard syntactic creativity as the "essential and defining characteristic of human language." However, as Katz (1976) noted, this does not constitute the unique feature of natural language since it is a consequence of any system's having a recursive structure. To dismiss primate language studies as failing to demonstrate syntactic ability is at best presumptuous. One presumption is that we have an acceptable theory of syntax ready at hand.

Physiological Structure

Perhaps there is a more important sense of structure that differentiates primates from humans. This is not to say simply that primates lack the human vocalization capacity but rather that they lack the requisite neural structure for language (Lenneberg, 1971). A good example of this sort of structuralist reasoning is Robinson's (1976) claim that animal vocalizations originate in the limbic system, which governs emotional responses, whereas human speech originates in the neocortex. (The notable exception in humans is that vulgar and profane speech seems to be limbic in origin.) Hence, animal vocalization sits on the emotional side of the neural and evolutionary fence, whereas human speech is, for the most part, on the reason side.

This type of physiological conjecture can be problematic. First, there are historical cases in which these pronouncements have been undermined. As Griffin notes (Chapter 11, this volume), lateralization, once claimed to be unique to humans, was found in songbirds for vocalization, while many otherwise normal humans do not have larger speech control centers in one hemisphere. Second, functional localization is empirically and conceptually risky. Empirically, these regions are not well delineated; Bogen and Bogen's (1976) confusion over the many different proclaimed locations of Wernicke's region, which supposedly controls some aspects of verbal comprehension, is a case in point. Furthermore, it is not clear in what sense any one area of the brain has autonomous control over any one function.

Computers, Animals, and Theories of Language

Let us, however, grant the truth of the species-specific thesis for physiological language structure. Even if primates have totally different physical structures than humans, language is still possible for these systems. This would make the failure of early attempts to teach chimpanzees vocal language irrelevant to the language controversy (Kellogg, 1980). The best evidence we have for the functional irrelevancy of physical structure is an artificial system, the computer understanding of natural language. Computers have radically different underlying physical structure compared to primates and humans, and yet in terms of proven linguistic ability, Winograd's program SHRDLU or Schank's MARGIE far outdistance any nonhuman primate's proven performances. For example, these machines have no difficulty in generating novel syntactic structures, presuming we know what these are.

In fact, methodologically, computers are much better instruments to use in constructing theories of language than primates—if that is the goal of these research paradigms. Since the debate over primate language appears to presuppose and/or to need a theory of language, this research goal seems quite reasonable. Some advantages of artificial intelligence research are the following:

(1) *Experimental control.* Artificial intelligence researchers have almost complete control over their experimental subjects. You can literally creat language ex nihilo. Working with computers would supposedly be far less constraining than working with primates.

(2) *Top-down Strategy.* The top-down theorist "begins with a more abstract decomposition of the highest levels of psychological organization, and hopes to analyze these into more and more detailed smaller systems or processes until finally one arrives at elements familar to the biologists" (Dennett, 1978, p. 110). Much of cognitive psychology that employs this strategy can be viewed as a reaction to the failure of traditional bottom-up strategies, such as behaviorism, which begin with some well-defined psychological atom (stimulus–response) and build up to more complex levels of organization. With the top-down approach researchers ask the following: How could any system (with features A, B, C, . . .) possibly accomplish X? In other words, as psychologists we want to find out how something is done in any system. We do not even know how language is possible, to say nothing of whether it is possible in primates.

(3) *Integrative.* Transformational grammar is very much of a faculty psychology with a fairly distinct faculty set aside, not only for language but also for syntax, and the theory's basic structure makes it extremely difficult to graft any of these faculties onto others in the whole cognitive system. One of the lessons learned very quickly from artificial intelligence research was the necessity for a more integrative system that, for example, included world knowledge for understanding natural language. It should be noted, however, that the nature of this integration in artificial intelligence research is largely cognitive.

Even though one may accept artificial intelligence as more integrative than transformational grammar, it still seems more difficult to make an analogous case for artificial intelligence over live organisms. After all, nonhuman primates with intact limbic systems, motor systems, etc., are more like us than machines could ever be. Accordingly, animals, at least of the nonhuman primate variety, are better models, on this account, for linguistic investigation. Nevertheless, there is a central ingredient missing or at least difficult to find in the primate model but found as a critical link in the artificial intelligence model which we have already mentioned—world knowledge. Without world knowledge early word-by-word machine translators had difficulty with sentences such as "The pen is in the box" and "The box is in the pen." Although recent work in artificial intelligence has not cured that (see Boden, 1977), it has certainly brought us a long way toward integrating world knowledge into language understanding. Could not a similar argument be made for primate understanding? Although a case could be made, and I would be the last to deny belief and intention ascriptions to primates, an equally strong case could be made that primate world knowledge is quite different from human world knowledge. There is no reason to believe that the knowledge gap across cultures is any less than that between species; hence, the difficulty in deciphering primate semantics. However, the important point for our purposes is that this epistemological component can be manipulated and the integrative nature of language more assured with artificial intelligence models than with animal ones.

As we shall see, this discussion of computer models in linguistics provides a convenient bridge to some important ethical issues, but before addressing these some general observations about the critiques of primate language studies need to be made. First, although this need not be the case, many of the criticisms of primate language research assume some other adequate linguistic theory when actually the entire field of linguistics is in a state of turmoil and disarray. There are no really acceptable standards of where we should even begin searching for a theory. Second, the criticisms often assume that there is a laundry list of necessary and/or sufficient conditions adequate for defining language as such. While this is an overly restrictive demand, not unlike those imposed by behaviorism, there is something of a cart–horse priority problem here as well. Part of the task of primate language studies is to uncover features of language and not have them imposed ex cathedra.

Finally, there is a tiring and frustrating game being played with these critiques which may be of interest to the sociology of science but is of questionable value for the advancement of knowledge. I suppose there are advantages to having the analogue to what artificial intelligence workers refer to as the "Dreyfus affair" in these studies. Dreyfus (1972) proposed a list of "cannots" for computers that have in some rare instances been taken as challenges by what Weizenbaum (1976) referred to as the "artificial intelligentsia." All too often, however, these critiques serve more as evidence for how little understanding there is between various research factions rather than as friendly goads.

At the very least, we have shown that the science of nonhuman primate language study cannot be summarily dismissed. While the primate language researchers have not conclusively demonstrated semantic and syntactic skills, it is very unclear, partially because of the state of the art, just what would constitute sufficient proof to satisfy the skeptic, who, in some cases, has resorted to impossible requirements such as a physiological structure similar to humans. Whatever our predilections, the primate language hypotheses are viable and respectable. However, just because these studies are viable does not mean that they should be conducted.

Ethics of Primate Language Research

If, as I have argued, computer models are viable alternatives to nonhuman primate models, then we are confronted with not only the aforementioned methodological problems but with some ethical ones as well. One ethical charge can be put rather succinctly: If suitable alternatives to live nonhuman animals can be found for scientific research, then those alternatives ought to be used instead of animals. Let us say that the Ames test for carcinogens, which uses bacteria, is a suitable replacement for using live animals, or that computer simulation studies provide an alternative to the LD_{50} (lethal dosage for 50% of a population of experimental animals) test. Would we not then have a moral obligation to use these rather than animals? Is the primate case any different?

Animal rights advocates and others have been slow to criticize, to say nothing of condemn, the primate studies. There is a tactical reason for this: with so many blatantly torturous experiments being conducted, it would be unwise to attack primate research that is at least compatible with a humane ethic. After all, unlike animals in an LD_{50} test, the primates are not being systematically made to suffer. Nevertheless, I think that there is a deeper, more fundamental reason for this avoidance behavior on the part of animal liberation advocates. Apparently innocuous practices can often conceal underlying problems. In making a case for animal liberation, Singer (1975) extended the notion of suffering beyond pain. Chickens bred under factory farm conditions may not feel pain, but their crowding may still constitute suffering. While the primates in language experiments are certainly treated far better than agribusiness animals, suffering may be inflicted in subtler ways.

The problems begin with procurement. According to a 1975 Institute of Laboratory Animal Resources survey, to procure 65,000 primates for research another 85,000 are sacrificed (Pratt, 1976). (For every chimpanzee that reaches a laboratory, 6 or 7 others die from trapping, holding by dealers, transport, and quarantine. In 1973 primate births in the United States represented less than 5% of the annual demand, and 54 species of primates, which constitutes 55% of all primate species, are severely threatened.) Captive breeding may be a lesser harm, but nonetheless a harm. Without strong countervailing reasons, the mere caging of an animal would constitute inflicting suffering. Restricting freedom unless it is for the organism's own good is morally questionable. Furthermore, severing a young infant from her or his mother is deplorable for primate and human alike. This is the treatment afforded many of the primates used in these studies.

The picture we get is a far cry from the cute-chimp-in-diapers scenes depicted in films about these primates. As we have seen, harm is perpetrated on the primates in various ways: through the procurement process (killing other primates, taking the primates out of their natural habitat), and in the laboratory setting (restricting freedom through caging and denying social interaction). If alternative aproaches to this research, such as using computers and other means, are available, why is there any need and any justification for intentionally inflicting these harms?

It is possible that these misgivings miss the whole point—at least the whole ethical point—of this research. By demonstrating that primates have language, do not the positive results of this research raise the ethical status of primates? An affirmative answer to this places the primate researcher in an interesting bind. The more successful the research findings, that is, the more humanlike with respect to language primates are found to be, the more ethically questionable these undertakings become. If primates have a language capacity, then experimenting on them should be addressed within the context of informed consent before the university committee on human, not animal experimentation; yet, despite a language capacity or even performance, how could primates ever be said to be fully informed about the experiment? To be fully informed on these matters requires background knowledge similar to our own. If we were to act on the primate's behalf as an adult might do for a child, a strong case could be made for not allowing the experiment to proceed because of the potential harms previously mentioned.

All of this presupposes a certain view of the relationship between language and morality: rights and duties have meaning only when they are expressible in language. Accepting the moral significance of language capacity downgrades the ethical status of lower species while improving that of higher ones such as primates. However, why should our ability to use language place us in any privileged moral position?

> Educated chimps would not form, along with man, an exploitive elite, exalted above all other life-forms as subject above object. The only intelligible arrangement is to regard *all* animals as subjects of some kind, though with a life that varies greatly in its kind and degree of complexity. (Midgley, 1978, p. 225)

The morally relevant question is not whether the primates can attain a certain linquistic level of performance but whether they have the capacity to suffer (Singer, 1975). There is little dissent to the claim that the great apes used in these language studies do have the capacity to suffer. Our moral action, then, does not depend on the results of these language studies, but rather on our ability to admit that there is no overwhelming need to perpetuate suffering, however innocuous it might first appear, on our fellow primates.

However, we may be comparing the incomparable. Perhaps artificial intelligence is not a viable alternative to primate language research since the aims of the respective research programs are quite different. While the difference is one of degree, artificial intelligence is more theoretically and the primate research more descriptively oriented. If so, then the primate research should be more properly compared to language acquisition studies in children.

Even here, however, a case can be made for preferring to conduct experiments on children rather than on primates. First of all, it is the acquisition skills of humans in which we are primarily interested when experimenting on primates. Hence, why not avoid extrapolation problems and go right to the primary source? Second, almost all children acquire language naturally, whereas we are clearly imposing an unnatural behavior on the primates. Cases that more closely parallel the primate case are those rare exceptions of children who have been deprived of the opportunity to learn a language. Still the cases are different. The deprivation has set unnatural conditions for the child, whereas the experimenter is setting the unnatural conditions for the primate.

This is not to say that nature is benign nor that the natural is ethically preferable; rather, it is to say that humans, unlike much of nature, can intentionally avoid inflicting harm. Imposing the unnatural ought to be avoided unless it can be shown that it is in the animals' best interest to have something unnatural imposed on them. What could possibly be ethically objectionable about trying to make another animal wiser?

> Let's suppose, as I think is the case with at least Premack and the Gardners, that the experimenter truly believes that learning the symbolic system will provide the chimp with new cognitive resources and not deprive her of any she already has. In other words learning the system will make the chimp wiser and thereby enrich her range of experiences. Now it is true that she must suffer the many indignities inherent in being a student, but this is no more than we inflict on our own children "for their own good." (Gopnik, 1981b)

However alluring this position might be, it raises some fundamental questions about the nature of language and the nature of science. These questions will be raised within the context of two analogies designed to show that we do not even really know what is in the best interest of our own species when it comes to language; how can we claim to know what is the primate's best interest?

The first analogy involves developmental problems in education. To put the problem crudely: in our culture if you do not have certain *kinds* of language skills, then you are in serious educational trouble. Children with mixed brain dominance, who learn language through sight and not phonetics, are often labeled slow learners, as are those who learn better through tactile modality than through the "normal" channels of sight and hearing.

Two problems are raised by these examples. First of all, they help illustrate that language is a more complex and more integrated phenomena than is assumed by most current theoretical conceptions of it. Most linguistic theories treat language as if it were solely a matter of manipulating discrete symbols within a purely cognitive domain. Separating language from context seems to be a highly questionable research practice.

> In most situations it is not a single signal that passes from one animal to another but a whole complex of them, visual, auditory, tactile, and sometimes olfactory. There can be little doubt that the structure of individual signals is very much affected by this incorporation in a whole matrix of other signals. (Marler, 1965, p. 583)

To ignore the multisensory modality of language raises serious methodological questions. To impose a particular conception of language that ignores this in children and primates similarly raises serious ethical questions. Imposing a narrow conception of language indicates a lack of sensitivity to and respect for the diversity within our own species. If this practice is unjustified when educating members of our own species, it is even more questionable when considering the training of primates.

Second, by looking at the educational process some of the problems and pitfalls of overemphasizing language become apparent. For example, education pays scant attention to the power of visualization. People who are good visualizers but poor verbalizers are put to a disadvantage. In keeping with this attitude, primates are not studied for their creative, imaginative abilities. Our language orientation may even mislead us in characterizing a highly cognitive internal state such as internal world representation.

> Beyond question the chimpanzee manipulates internal *world representations* of some highly sophisticated kind. But for us to represent his representings on the overtly *linguistic* model that dominates (almost constitutes) our own theory of mind is to be insupportably parochial again. ... Granted, human language is so far the only model for systematic world representation we possess; but that is a defect in our intellectual situation that desperately needs remedy, arguable as it is that the linguistic model is inadequate even to represent our own cognitive activities. (Churchland & Churchland, 1978, p. 565)

Language, then, may not be the primary avenue to the mind—our own or that of other species. Moreover, language may not be the primary conveyor of meaning. As Fouts (1978) noted, it is estimated that 55%–80% of the meaning generated in a two-person conversation stems from nonverbal communication. To overemphasize language is methodologically and ethically suspect. Much more could be learned from primates in other ways. If overemphasizing linguistic skills can be harmful to humans, it is even more harmful to primates.

The other analogy consists in imposing white middle-class English on minority groups such as blacks. Black English can be viewed as a separate language that helps blacks achieve social cohesion and identity. While one can argue that imposing white English on blacks is in the black's best interest, the case is not altogether closed. Assimilation of one culture into another through linguistic imperialism can be highly detrimental to the members of that group. If it is not clear that imposing white English is in the best interests of blacks, then it is certainly much less clear than imposing the same language on primates is in their best interest.

Imposing white English on blacks and primates points to a lack of respect for both. For the question, What can be learned from black English?, is hardly ever asked. Seldom is any attempt made by whites to learn black English. Likewise, little effort is made to determine what we can learn from the primates not through laboratory conditions set by us, but through observing the primates in their natural habitat. It always amazed me that Lilly (1975) spent so much time coaxing dolphins into imitating human nonsense sounds and so little time imitating the richly complex communication system of the dolphins.

Field studies of humans and ethological studies of animals are methodologically and ethically superior to the laboratory setting imposed on both humans and animals. Controlled experimentation in a laboratory is regarded as being a sine qua non of scientific methodology; this position is adamantly taken by most social scientists. Adhering to this structure, however, would mean discarding disciplines such as astronomy, which depends heavily upon natural observations.

Although arguing the case fully would take us beyond the scope of this discussion, the methodological superiority of field studies can be evidenced by the fact that field, unlike laboratory, studies take much fuller account of context. When examining communication, context is crucial. Communication occurs in a historical–social context. To isolate one aspect of communication, such as symbol manipulation, to the exclusion of other aspects of the complex, such as the affective domain, is to distort the communication process. Field studies enable the researcher to describe more accurately what is actually happening and do not lend themselves as easily as laboratory studies do to finding what the researcher wants to find.

Field studies also provide additional ethical benefits. Basically, the subject is studied on her or his own terms with a minimum of restrictions being imposed by the researcher. This approach develops a new respect for the subject by treating the subject as fellow subject and not solely as object.

Viable alternatives to primate language research are readily available. Computer models, human subjects, and field studies all provide alternatives that inflict far less suffering than the primate laboratory studies. Not only do they have this ethical advantage, but also a strong case can be made for the methodological advantage of each over the primate experiments. Adopting any of these does not have the associated disadvantage of our losing valuable information by not doing the primate research.

Conclusions

Methodologically, the primate language studies constitute a research program that can be defended against critics' charges. It is these charges that often make unwarranted assumptions. A degree of experimental control is demanded of the primate experiments that cannot even be met by the language acquisition studies of children. For example, were all subliminal cues from experimenters and the social environment controlled for in the latter experiments? Furthermore, the critics assume particular theories of language that are themselves debatable. Finally, some critics automatically rule out the results of the primate research. For example, transformational grammar adherents assume that language is the sole domain of humans. Physiological structure and other features have been proposed as necessary conditions for language, but these are unwarranted. For these and other reasons it is not justified to dismiss the primate research program as unscientific or hopeless. Even the results uncovered so far are at the very least controversial because of the state of the language studies.

Defending the primate studies in this way, however, does not mean that they are methodologically acceptable. In fact, I have argued that a number of different approaches are to be preferred, methodologically, over the laboratory primate studies. If we are interested in constructing a cognitive theory of language, then a strong case can be made for turning to computers instead of primates. If our concerns are more oriented toward language acquisition, then we should turn to humans instead of primates. Finally, if our aim is to uncover the communication process in all its contextual richness, then field instead of laboratory studies are preferable.

However, this is not simply a matter of methodological preference. Each methodological choice has important ethical ramifications. Overall, the availability of these different approaches means that there are alternatives to conducting primate language research in laboratory settings. We ought to avail ourselves of these alternatives since each of them inflicts less suffering than the primate experiments.

The cry "we're doing it for their own good" will not suffice since it is most often the case that we cannot even determine what is our own good in these matters. In fact, it seems that it is not in our best interest either to impose language in some exclusive, domineering fashion on others or to impose a particular language or conception of language on others.

Acknowledgments

I would like to thank Professors M. Gopnik, Martin Ringle, and Hugh Wilder for their comments on an earlier draft. Part of the paper was written while I was at the Center for the Study of Values at the University of Delaware. I would like to thank the Center and its sponsor, the Exxon Foundation, for their support.

References

Boden, M. *Artificial intelligence and natural man*. New York: Basic Books, 1979.

Bogen, J. E., & Bogen, G. M. Wernicke's region—Where is it? In S. Harnad, H. D. Steklis, & J. Lancaster, (Eds.), *Origins and evolution of language and speech*. New York: New York Academy of Sciences, 1976, pp. 834–843.

Brown, R. *Psycholinguistics*. New York: Free Press, 1970.

Chomsky, N. *Cartesian linguistics*. New York: Harper & Row, 1966.

Churchland, P. S., & Churchland, P. M. Internal states and cognitive theories. *The Behaviorial and Brain Sciences*, 1978, *4*, 565–566.

Dennett, D. *Brainstorms*. Montgomery, Vt.: Bradford Books, 1978.

Dreyfus, H. *What computers can't do*. New York: Harper & Row, 1972.

Fodor, J. A., Bever, T. G., & Garrett, M. F. *The psychology of language: An introduction to psycholinguistics and generative grammar*. New York: McGraw-Hill, 1974.

Fouts, R. S. Sign language in chimpanzees. In F. C. C. Peng (Ed.), *Sign language and language acquisition in man and ape: New dimensions in comparative psycholinguistics*. Boulder, Colo.: Westview Press, 1978, pp. 121–136.

Gardner, R. A., & Gardner, B. T. Teaching sign language to a chimpanzee. *Science*, 1969, *165*, 644–672.

Gopnik, M. *The ape language controversy*. Unpublished manuscript, McGill University, 1981. (a)

Gopnik, M. Personal communication, 1981. (b)

Griffin, D. R. *The question of animal awareness*. New York: Rockefeller University Press, 1976, p. 74.

Hockett, C. F., & Altman, S. A. A note on design features. In T. A. Sebeok (Ed.), *Animal communication*. Bloomington, Ind.: Indiana University Press, 1968, pp. 61–72.

Katz, J. J. A hypothesis about the uniqueness of natural language. In S. Harnad, H. D. Steklis, & J. Lancaster (Eds.), *Origins and evolution of language and speech*. New York: New York Academy of Sciences, 1976, pp. 33–41.

Kellogg, W. N. Communication and language in the home-raised chimpanzee. In T. A. Sebeok & J. Umiker-Sebeok (Eds.), *Speaking of apes*. New York: Plenum Press, 1980, pp. 61–70.

Lenneberg, E. H. Of language, knowledge, apes, and brains. *Journal of Psycholinguistic Research*, 1971, *1*, 1–29.

Lilly, J. C. *Lilly on dolphins*. Garden City, N. Y.: Anchor Press, 1975.

Marler, P. Communication in monkeys and apes. In I. DeVore (Ed.), *Primate Behavior*. New York: Holt, Rinehart & Winston, 1965, pp. 544–585.

Menzel, E. W., & Johnson, M. K. Should mentalistic concepts be defended or assumed? *The Behavioral and Brain Sciences*, 1978, *4*, 586-587.

Midgley, M. *Beast and man*. Ithaca, N.Y.: Cornell University Press, 1978.

Patterson, F., & Linden, E. *The education of Koko*. New York: Holt, Rinehart & Winston, 1981.

Pratt, D. *Painful experiments on animals*. New York: Argus, 1976.

Pylyshyn, Z. M. When is attribution of beliefs justified? *The Behavioral and Brain Sciences*, 1978, *4*, 592–593.

Robinson, B. W. Limbic influences on human speech. In S. Harnad, H. D. Steklis, & J. Lancaster (Eds.), *Origins and evolution of language and speech*. New York: New York Academy of Sciences, 1976, 761–771.

Savage-Rumbaugh, E. S., Rumbaugh, D. M., & Boysen, S. Linguistically mediated tool use and exchange by chimpanzees. *The Behavioral and Brain Sciences*, 1978, *4*, 539–554.

Singer, P. *Animal liberation*. New York: New York Review of Books, 1975.

Umiker-Sebeok, J., & Sebeok, T. A. Questioning apes. In T. A. Sebeok & J. Umiker-Sebeok (Eds.), *Speaking of apes*. New York: Plenum Press, 1980, pp. 1–59.

Weizenbaum, J. *Computer power and human reason*. San Francisco: Freeman, 1976.

CHAPTER 7

Ethics, Animals, and Language

Martin Benjamin

Are there any ethical restrictions on the way in which human beings may use and treat nonhuman animals? If so, what are they and how are they to be justified? I first review three standard responses to these questions and briefly indicate why none of them is entirely satisfactory. Then I identify what I take to be the kernel of truth in each of the three responses and attempt to blend them into a fourth, more adequate position. In so doing, I hope to identify the significance of language in determining ethical restrictions on our use and treatment of nonhuman animals.

Three Standard Positions

Historically, western philosophers have responded to questions about the nature and extent of ethical restrictions on the human use and treatment of nonhuman animals in three ways. First, those who hold what I label "indirect obligation" theories maintain that ethical restrictions on the use and treatment of animals can be justified *only* if they can be derived from direct obligations to human beings. The second type of response, which I label "no obligation" theories, holds that there are no restrictions whatever on what humans may do to other animals. The third type of response, which I label "direct obligation" theories, maintains that ethical restrictions on the use and treatment of animals can sometimes be justified solely for the sake of animals themselves.

Indirect Obligation

Among the most noted philosophers in the Western tradition, St. Thomas Aquinas (1225-1274) and Immanuel Kant (1724-1804) have acknowledged restrictions on human conduct with regard to the use and treatment of non-human animals, but these restrictions are, in their view, ultimately grounded upon obligations to other human beings. Blending views that can be traced both to the *Bible* and Aristotle, Aquinas (1265-1273/1918) held a hierarchial or means-end view of the relationship between plants, animals, and humans, respectively:

> There is no sin in using a thing for the purpose for which it is. Now the order of things is such that the imperfect are for the perfect . . . things, like plants which merely have life, are all alike for animals, and all animals are for man. Wherefore it is not unlawful if men use plants for the good of animals and animals for the good of man, as the Philosopher states (*Politics,* i, 3).
>   Now the most necessary use would seem to consist in the fact that animals use plants, and men use animals, for food, and this cannot be done unless these be deprived of life, wherefore it is lawful both to take life from plants for the use of animals, and from animals for the use of men. In fact this is in keeping with the commandment of God himself (*Genesis* i, 29, 30 and *Genesis* ix, 3). (quoted in Regan & Singer, 1976, p. 119)

Nevertheless, it does not follow, for Aquinas, that one can do anything to an animal. For example, one is still prohibited from killing another person's ox: "He that kills another's ox, sins, not through killing the ox, but through injuring another man in his property. Wherefore this is not a species of the sin of murder but of the sin of theft or robbery." There may even be similarly *indirect* grounds for not harming animals who are no one's property. Thus, Aquinas (1259-1264/ 1928) explained,

> if any passages of Holy Writ seem to forbid us to be cruel to dumb animals, for instance to kill a bird with its young: this is either to remove man's thoughts from being cruel to other men, and lest through being cruel to animals one become cruel to human beings: or because injury to an animal leads to the temporal hurt of man, either of the doer of the deed, or of another. (quoted in Regan & Singer, 1976, p. 59)

   Kant (1963), too, held that insofar as humans are obligated to restrain themselves in their dealings with animals, it is due to their obligations to other humans. Thus,

> so far as animals are concerned, we have no direct duties. Animals are not self-conscious and are there merely as a means to an end. That end is man . . . Our duties towards animals are merely indirect duties towards humanity. Animal nature has analogies to human nature, and by doing our duties to animals in respect of manifestations of human nature, we indirectly do our duty to humanity . . . . If . . . any acts of animals are analogous to human acts and spring from the same principles, we have duties towards the animals because thus we cultivate the same duties towards human beings. If a man shoots his dog because the animal is no longer capable of service, he does not fail in his duty to the dog, for the dog cannot judge, but his act is inhuman and damages in itself that humanity which it is his duty to show towards mankind. If he is not to

stifle his human feelings, he must practice kindness towards animals, for he who is cruel to animals becomes hard also in his dealings with men. (quoted in Regan & Singer, 1976, p. 122)

Thus Aquinas and Kant both hold what I have labeled "indirect obligation" theories with regard to ethical restrictions on the use and treatment of animals. Although they agree that we have obligations *with regard* to animals, these obligations are *not,* at bottom, *owed to* the animals themselves but rather they are owed to other human beings.

There are, nonetheless, significant problems with both Aquinas's and Kant's positions, at least in their present forms. First, insofar as Aquinas assumes that it is necessary for humans to use animals for food and thus to deprive them of life, his position must be reconsidered in the light of modern knowledge about nutrition. It has been maintained, for example, that a perfectly nutritious diet may require little or no deprivation of animal life and, even if it does, that the average American consumes twice as much animal protein as his or her body can possibly use (Lappé, 1975). Insofar as we continue to consume large quantities of animal foodstuff requiring pain and deprivation of life, then we do so not so much to serve vital nutritional demands, but rather to indulge our acquired tastes. Second, insofar as Aquinas's view is based upon a hierarchial worldview and assumes that those lower in the order or less perfect are to serve the good of those higher or more perfect, it is open to a serious theoretical objection. It is, unfortunately, not difficult to imagine a group of beings—perhaps from another part of the universe—who are more rational and more powerful than we. Assuming that such beings are more perfect than we are, it seems to follow, if we adopt the principles underlying Aquinas's view, that we ought to acquiesce in their using us for whichever of their purposes they fancy we would serve. Do we want to agree with the rightness of this? If we take Aquinas's view, would we have any grounds on which to disagree?

As for Kant's view, the main difficulties have to do first with his emphasis on self-consciousness as a condition for being the object of a direct obligation, and second with his assumption that all and only human beings are self-conscious. I will postpone consideration of the first difficulty until later. For the moment, let me simply develop the second. Even supposing that being self-conscious is a necessary condition for being the object of a direct obligation, it does not follow either that *all* human beings are the objects of direct obligations or that *no* animal can be the object of such an obligation. First, advances in medical knowledge, techniques, and technology have, among other things, preserved and prolonged the lives of a number of human beings who are severely retarded or otherwise mentally impaired due to illness or accident, or irreversibly comatose (e.g., Karen Ann Quinlan). In our day, then, if not in Kant's, one cannot assume that all human beings are self-conscious. Second, some contemporary researchers have suggested that at least some nonhuman animals have a capacity for becoming self-conscious that has, until recently, been undetected or ignored by humans. Thus, even if we follow Kant and accept self-consciousness as a condition for being the object of direct obligations, it does not follow that *all* and *only* humans satisfy this condition. Some humans, it may turn out, will not be the objects of direct obligations and some animals will.

No Obligation

If animals are not conscious—that is, if they are not sentient and have no capacity for pleasure, pain, or any other mental states—they may not even be the objects of indirect obligations. Insofar as Aquinas says that it is possible to be "cruel to dumb animals" and Kant says that "he who is cruel to animals becomes hard in his dealings with men," each presupposes that animals, unlike plants and machines, are sentient and are thereby capable of sensation and consciousness. Thus it is surprising to find René Descartes (1596-1650), a reknowned philosopher, mathematician, and scientist, comparing animals to machines. Nonetheless, this is just what he did in his influential *Discourse on Method* when he compared machines made by the hand of man with human and nonhuman animal bodies made by the hand of God: "From this aspect the body is regarded as a machine which, having been made by the hands of God, is incomparably better arranged, and possesses in itself movements which are much more admirable than any of those which can be invented by man" (Descartes, 1637/1968; quoted in Regan & Singer, 1976, p. 61). Living *human* bodies were, for Descartes, distinguished from living *animal* bodies by the presence of an immortal soul, which was a necessary condition for mental experiences. Without a soul, a living biological body was a natural automation, "much more splendid," but in kind no different from those produced by humans.

For Descartes (1637/1968), the criterion for distinguishing those living bodies which were ensouled from those which were not was the capacity to use language. The former, he believed, included all and only human beings. Among humans, he maintained,

> there are none so depraved and stupid, without even exempting idiots, that they cannot arrange different words together, forming of them a statement by which they make known their thoughts; while on the other hand, there is no other animal, however perfect and fortunately circumstanced it may be, which can do the same. (quoted in Regan & Singer, 1976, p. 61)

Insofar as nonhuman animals do appear to do some things better than we do, Descartes (1637/1968) added, "it is nature which acts in them according to the disposition of their organs, just as a clock, which is only composed of wheels and weights is able to tell the hours and measure the time more correctly than we can do with all our wisdom" (quoted in Regan & Singer, 1976, p. 62). As for the ethical implications of his view, Descartes (1649/1970), in a letter to Henry More, noted that his "opinion is not so much cruel to animals as indulgent to men . . . since it absolves them from the suspicion of crime when they eat or kill animals" (quoted in Regan & Singer; 1976, p. 66).

Insofar as Descartes' position presupposes that all and only human beings have the capacity to use language, it is open to the same sort of criticisms and objections that we raised against Kant. That is, advances in medicine are providing more nonlinguistic humans and advances in science are suggesting that at least some nonhuman animals have more linguistic facility or capacity than we previously supposed. Moreover, even if Descartes were correct in believing that the capacity to use language is uniquely human, why should this, rather than the capacity to feel pain and experience distress, be the principal criterion for determining the nature and extent of ethical restrictions on the use and treatment of animals? It is this objection that sets the stage for positions which hold that humans have direct obligations to at least some animals.

Direct Obligation

Jeremy Bentham (1748-1832), the father of modern utilitarianism, held that pain and pleasure were what governed behavior and that any ethical system that was founded on anything but maximizing the net balance of pleasure over pain dealt in "sounds instead of sense, in caprice instead of reason, in darkness instead of light." Every action, for Bentham, was to be assessed in terms of its likelihood of maximizing the net balance of happiness. However, he noted, if the capacity to experience pleasure and pain was what qualified one to be taken into account in estimating the effects of various courses of action, then nonhuman as well as human animals would have to be taken into account insofar as they, too, have the capacity to experience pleasure and pain. Thus, for Bentham (1789), it is sentience, or the capacity for pleasure and pain, that determines whether a being qualifies for moral consideration.

> What else is it that should trace the insuperable line? Is it the faculty of reason, or perhaps the faculty of discourse? But a full-grown horse or dog is beyond comparison a more rational, as well as a more conversable animal than an infant of a day, or a week, or even a month, old. But suppose the case were otherwise, what would it avail? The question is not, Can they *reason*? nor, Can they *talk*? but, *Can they suffer*? (quoted in Regan & Singer, 1976, p. 130)

The question now is, What grounds do we have to believe that animals *can* suffer, feel pain, or experience distress? If a being lacks the capacity to convey its suffering, pain, or distress linguistically, how do we know that it actually has such experiences and isn't a rather splendid automaton going through the motions?

In response to such skepticism, one holding a utilitarian direct obligation theory must show why he or she believes that nonhuman animals are conscious. There are a number of ways one might go about this. First, one could stress behavioral similarities between human and nonhuman animals in their respective responses to certain standard pain- and pleasure-producing stimuli. Comparing the behavior of nonhuman animals with human infants would be especially forceful here. Second, we could stress relevant neurophysiological similarities between humans and nonhumans. After making these comparisons, we may then be inclined to agree with Sergeant (1969, p. 72) when he claimed the following:

> Every particle of factual evidence supports the contention that the higher mammalian vertebrates experience pain sensations at least as acute as our own. To say that they feel less because they are lower animals is an absurdity; it can easily be shown that many of their senses are far more acute than ours—visual acuity in certain birds, hearing in most wild animals, and touch in others; these animals depend more than we do on the sharpest possible awareness of a hostile environment. (quoted by Singer, 1980, p. 225)

If Sergeant is correct in this, at least some animals are conscious and, hence, on utilitarian grounds, qualify as the objects of direct obligation.

There are, nonetheless, significant limitations to this view. First, although utilitarianism takes nonhuman animals directly into account in determining ethical obligations, there is no guarantee that animals will, in fact, fare better in this view than they will in an indirect obligation view like that of Aquinas or Kant. Contemporary animal welfare advocates who find utilitarianism hospitable to their position have not fully appreciated utilitarianism's indifference to any outcome apart from the maximization of happiness. Thus, for example, on utilitarian grounds, a policy that causes a great amount of pain to animals but also causes an even greater amount of offsetting pleasure to humans would appear to be ethically justified. Second, one who adopts utilitarianism because it takes direct account of animal suffering must recognize all of its implications. One of the standard objections to utilitarianism is that it seems, on the face of it, more suited to animals than it is to human beings. Thus Bentham's version was initially caricatured as philosophy for swine because it seemed to imply that it was better to be a satisfied pig than a dissatisfied human; or better to be a fool satisfied than Socrates dissatisfied.

A Fourth Position

Although none of the positions we have examined is entirely satisfactory, each, I believe, has something to recommend it. Indirect obligation theories are correct to stress the difference between what I will call "simple consciousness" and "reflective consciousness," but they have not adequately characterized the difference nor have they fully appreciated its ethical significance. No obligation theories, at least that of Descartes, are correct in emphasizing the relationship between the use of language and the development of reflective consciousness. Finally, direct obligation theories are correct in noting that the possession of simple consciousness (or sentience) in human or nonhuman animals is, by itself, sufficient to give them independent standing in the ethical deliberations of beings who are reflectively conscious. Each of these fundamental insights may be integrated into a fourth, more adequate position.

The fundamental insight of indirect obligation theories is their recognition of a difference between simple and reflective consciousness. Beings having only simple consciousness can experience pain, have desires, and make choices. However, they are not capable of reflecting upon their experiences, desires, and choices and altering their behavior as a result of such self-conscious evalution and deliberation. Beings who can do this I will, following John Locke (1632-1704), label "persons." A person, in Locke's (1690/1961, p. 281) view, is "a thinking intelligent being that has reason and reflection and can consider itself as itself, the same thinking thing, in different times and places." Although they were mistaken in believing that the class of persons fully coincided with the class of human beings, indirect obligation theorists were correct to emphasize the special status of persons. For only persons are capable of tracing the consequences and implications of various courses of action and then deliberating and deciding to embark on one rather than another on grounds other than self-interest. To do this is part of what it means to have a morality and it is the capacity for taking the moral point of view (that is, voluntarily restricting one's appetite or desires for the sake of others) that gives the persons their special worth.

The fundamental insight of Descartes' no obligation theory was to recognize the connection between the development and exercise of personhood and the development and exercise of language. As Hampshire (1979, p. 42d) recently pointed out, although people often associate the use of language primarily with communication, "language's more distinctive and far-reaching power is to bring possibilities before the mind. Culture has its principal source in the use of the word 'if,' in counterfactual speculation." Only language, then, gives us the power to entertain complex unrealized possibilities. "The other principal gift of language to culture," Hampshire continued, "is the power to date, and hence to make arrangements for tomorrow and to regret yesterday." Thus a being cannot become a person and, in Locke's words, "consider itself as itself, the same thinking thing, in different times and places," without the use of language.

Finally, the fundamental insight of direct obligation theories was to note that one needn't be a person to be the object of a moral obligation. Simple consciousness or sentience is sufficient to entitle a being to be considered *for its own sake* in the ethical deliberations of persons. If, for example, the capacity to feel pain is sufficient ground for a prima facie obligation not to cause gratuitous pain to persons, why is it not also a sufficient ground for a similar obligation not to cause pain to beings having simple consciousness? With regard to the evil of avoidable and unjustifiable pain, the question is, as Bentham emphasized, not "Can they reason? nor, Can they talk? but, Can they suffer?"

Putting all of this together, we may say that persons, who are characterized as possessing reflective consciousness, may have a higher status than beings having only simple consciousness. Their special worth is a function of the extent to which they use language "to bring possibilities before the mind" and then restrain their more trivial desires for the sake of not harming others whom they recognize, from the moral point of view, as their equals in certain respects. Among the beings whose interests must be taken into account *for their own sake* in the moral deliberations of persons are beings possessing only simple consciousness. To the extent that persons reluctantly cause pain, suffering, and even death to beings possessing simple consciousness in order to meet *important needs,* what they do may be justified by appeal to their higher status or greater worth. However, to the extent that persons inflict avoidable pain and suffering on such beings merely to satisfy certain *trivial tastes or desires,* they pervert their greater capacities. In so doing, they ironically undermine their claim to higher status or worth and thereby weaken any justification they may have had for sacrificing beings having only simple consciousness for important ends.

Language, Personhood, and Nonhuman Primates

If this position, which I will label a "nonutilitarian direct obligation" theory, is correct, we may draw some interesting conclusions about the significance of language in determining ethical restrictions on our use and treatment of nonhuman animals. First, whether animals have, or can be trained to use, language (in some suitably rich sense) is not of crucial importance from an ethical point of view. To be taken seriously for their own sake it is enough, as Bentham emphasized, that animals are sentient. However, the fact that humans—or more precisely that large class of humans that also falls into the class of persons—have language is of central importance. The fact that persons can use language to trace the consequences and implications of various courses of action and to deliberate and decide to embark on one course of action rather than another is what provides both the basis for our *special worth* (thus allowing our vital interests to override those of nonpersons in cases of unavoidable conflict) and our *special obligations* (thus requiring us to restrain our more trivial tastes or desires if they would cause harm to nonpersons).

As suggested above, however, it is possible that some nonhuman animals—for example, nonhuman primates who are the subjects of various studies with regard to linguistic skills and acquisition—may qualify as persons. In order to determine if this is so, we must develop a deeper understanding of the relationship between language and personhood. Thus I want to conclude by identifying the sort of linguistic behavior that must be displayed by nonhuman primates before it can be said to have any *special* significance from an ethical point of view: that is, before such primates can enter into our ethical deliberations as (equal) persons.

First, we must acknowledge that any instance of a nonhuman primate's being observed or trained to use language in hitherto unsuspected ways is of *some* ethical significance. For if such uses of language reveal or help create various desires or appetites that can be satisfied or frustrated by what we do, we must take them into account in determining our conduct. Thus, insofar as the use or acquisition of language by nonhuman primates is related to an increase in their desires or appetites, it makes a difference from an ethical point of view. This is a difference of degree, however, and not of kind. The difference between a chimpanzee with an extensive set of desires made possible by and expressible in ASL and a chimpanzee without such an extensive set of desires is similar to the difference between a typical chimpanzee and a typical lower animal. The one has more desires than the other and thus there are more things that a person can do in the way of satisfying or frustrating its desires; but the difference is basically quantitative.

However, the difference between persons and sentient nonpersons—the difference between beings having reflective consciousness and beings having only simple consciousness—is a matter of kind and not of degree. What, then, is this difference and exactly what sort of linguistic behavior must we find in a nonhuman primate if we are to say that it is a person and not simply a nonperson with an exceptionally wide range of desires?

In order to answer this question it is helpful to follow Frankfurt (1971) in distinguishing "first-order" from "second-order" desires. The form of a first order desire is

(1) A wants$_1$ X

where A is the subject and X, the object, is some (positive or negative) action. Thus, for example, A may be a person or a nonperson, a human or a nonhuman, and X may be "to eat," "to run," "to avoid pain," "to hide," "to sleep," etc. The form of a second-order desire is

(2) A wants$_2$ to want$_1$ (or not to want$_1$) X

where A is once again the subject, but where the object of A's second-order desire is to have or not to have an effective first-order desire for X. Thus the object of a second-order desire is one of the subject's first-order desires.

For example, suppose that I have powerful first-order desires to smoke cigarettes. Having learned about the hazards of smoking, I then form a second-order desire to rid myself of my first-order desire to smoke or, at least, to order my first-order desires in such a way that my first-order desire to smoke is no longer effective. In so doing, I will, in Frankfurt's analysis, have effectively exercised my free will; that is, in this area of my life (at least) I will have brought it about that my effective first-order desire conforms to my second-order desire (or *volition*). With regard to smoking, then, my effective want$_1$ will be the want$_1$ that I want$_2$; or, in other words, I will have the will that I want$_2$ to have.[1]

Both persons and nonpersons, Frankfurt argued, are alike in having first-order desires, but only persons have the capacity for freedom of the will, where this notion is analyzed in terms of a subject's capacity to have the effective first-order desires that it (second-order) desires to have. As Frankfurt put it:

---

[1] Similarly, a person who learns that his or her first-order desires for large quantities of pale veal or richly marbled beef at the lowest possible cost are partly responsible for modes of livestock production involving considerable pain and suffering for calves and cattle may form a second-order desire to weaken or counterbalance such first-order desires. In *The Wizard of Oz* it is not his cowardice, but rather his second-order desires not to have or be moved by his first-order cowardly desires that makes the cowardly lion seem more like a person than a nonhuman animal.

Besides wanting and choosing and being moved *to do* this or that, men may also want to have (or not to have) certain desires and motives. They are capable of wanting to be different, in their preferences and purposes, from what they are. Many animals appear to have the capacity for what I . . . call "first order desires" or "desires of the first order," which are simply desires to do or not to do one thing or another. No animal other than man, however, appears to have the capacity for reflective self-evalutaion that is manifested in the formation of second-order desires. (Frankfurt, 1971, p. 7)[2]

Thus, "when a person acts, the desire by which he is moved is either the will he wants or a will he wants to be without" (Frankfurt, 1971, p. 14) When a non-person (human or nonhuman) acts, it is neither.

Supposing that this analysis of the freedom of the will in terms of a certain relationship between first- and second-order desires is correct and that some degree of freedom of the will in this sense is necessary to be a person, the question remains: What is the relationship between freedom of the will and the use of language?

A plausible answer to this question has recently been proposed by Dennett:

The "reflective self-evaluation" Frankfurt speaks of must be genuine self-consciousness, which is achieved only by adopting toward oneself the stance not only of a communicator but of . . . reason-asker and persuader. As Frankfurt points out, second-order desires are an empty notion unless one can *act* on them, and acting on a second-order desire must be logically distinct from acting on its first-order component. Acting on a second-order desire, doing something to bring it about that one acquires a first-order desire, is acting upon oneself just as one would act upon another person: one schools oneself, one offers persuasions, arguments, threats, bribes, in the hopes of inducing oneself to acquire the first-order desire. One's stance toward oneself *and access to oneself* in these cases is essentially the same as one's stance toward and access to another. One must *ask oneself* what one's desires, motives, reasons really are, and only if one can say, can become aware of one's desires, can one be in a position to change. Only here, I think, is it the case that the "order which is there" cannot be there unless it is there in episodes of conscious thought in a dialogue with oneself. (Dennett, 1976, p. 193)

---

[2] Note that although Frankfurt speaks here as if the distinction between persons and sentient nonpersons follows species lines, this is not his view. ". . . [T]he criteria for being a person do not serve primarily to distinguish the members of our own species from the members of other species. Rather, they are designed to capture those attributes which are the subject of our most humane concern with ourselves and the source of what we regard as most important and most problematical in our lives. Now these attributes would be of equal significance to us even if they were not in fact peculiar and common to the members of our own species. What interests us most in the human condition would not interest us less if it were also a feature of the condition of other creatures as well" (p. 6). Frankfurt uses the term "wanton" to refer to humans who may have first-order desires but no second-order volitions with regard to those desires.

I think it is safe to say, one cannot *ask* oneself what one's desires are, engage in a *dialogue* with oneself, and offer *persuasions, arguments, threats,* or *bribes* unless one is capable of doing so in language.

Thus in analyzing and cultivating the linguistic capacities of nonhuman primates the important thing is not so much whether such animals can (be taught to) communicate with us or each other, but rather whether they can (be taught to) communicate with themselves; that is, we must determine whether they can (be taught to) use language to "bring possibilities before the mind," use "the word 'if' in counterfactual speculation," and "make arrangements for tomorrow and . . . regret yesterday" (Hampshire, 1979, p. 42d). For only if they can use language in this way can they reflectively identify, evaluate, and either endorse or attempt to restructure various of their first-order desires. Only if they can do this have they the freedom of the will that is necessary for being a person. Finally, if and when some nonhuman primates or other animals are either trained or discovered to be persons, not only will we have new and more stringent obligations to them but, interestingly enough, they will have acquired some of the same obligations to us; for the same capacities that ground some of the rights of persons are, together with certain moral principles, sufficient for acquiring some of the responsibilities of persons.

Acknowledgment

The main ideas were first presented at an Interdisciplinary Symposium on the Question of Animal Consciousness at Michigan State University, April 5, 1980. Portions of this chapter have also appeared in T. Mappes & J. Zembaty (Eds.), *Social ethics: Morality and social policy* (2nd ed.). New York: McGraw-Hill, 1982.

References

Bentham, J. *The principles of morals and legislation.* 1789, Chap. XVII, Section 1.

Dennett, D. Conditions of personhood. In A. Oksenberg Rorty (Ed.), *The identities of persons.* Berkeley, Calif.: University of California Press, 1976.

Descartes, R. *Discourse on method.* In *Philosophical Works of Descartes* (Vol. 1) (E. S. Haldane & G. R. T. Ross, trans.). Cambridge, England: Cambridge University Press, 1637/1968.

Descartes, R. Letter to Henry More. In *Descartes: Philosophical Letters* (A. Kenny, Ed. and trans.). Oxford: Oxford University Press, 1649/1970.

Frankfurt, H. Freedom of the will and the concept of a person. *Journal of Philosophy,* 1971, *68,* 5-20.

Hampshire, S. Human nature. *New York Review of Books,* December 6, 1979, 42 c-d.

Kant, I. Duties to Animals and Spirits. In *Lectures on ethics* (L. Infield, trans.). New York: Harper & Row, 1963.

Lappé, F. M. Fantasies of famine. *Harper's, 250,* February 1975, 53.

Locke, J. *Essay concerning human understanding* (Vol.1). (J. Yolton, Ed.). London: Dent, 1961, Book II, Chap. XXVII.

Regan, T., & Singer, P. (Eds.). *Animal rights and human obligations.* Englewood Cliffs, N. J.: Prentice-Hall, 1976.

Sergeant, R. *The spectrum of pain.* London: Hart-Davis, 1969.

Singer, P. Animals and the value of life. In T. Regan (Ed.), *Matters of life and death.* New York: Random House, 1980, 218-258.

Thomas Aquinas, Saint. *Summa theologica* (English Dominican Fathers, trans.). New York: Benziger Brother, 1265-1273/1918, Part II, Question 64, Article 1.

Thomas Aquinas, Saint. *Summa contra gentiles* (English Dominican Fathers, trans.). New York: Benziger Brother, 1259-1264/1928, Third Book, Part II, Chap. CXII.

CHAPTER 8
Linguistic Innateness and Its Evidence

Margaret Atherton and Robert Schwartz

It has become increasingly popular of late to investigate the abilities of animals to learn to use a language. Not only have such phenomena as the dance of the bees, the songs of the whales, and the squeaks of the dolphin received close attention, but various attempts have been made to teach animals a natural language. Most notable of these perhaps have been the efforts to teach chimpanzees English. The Hayeses, for example, raised a chimpanzee, Viki, in a home environment similar to that of a preschool child, but Viki never learned to utter more than a few English words (Hayes & Nissen, 1971). The Gardners have increased the performance capabilities of their chimpanzee, Washoe, by teaching her sign language rather than spoken English, and Washoe is capable of understanding and generating a limited set of new sentences (Gardner & Gardner, 1971). Premack (1971) has pushed the linguistic competence of his chimpanzee, Sarah, even further. By using arbitrary plastic pieces for words, Premack claims to have taught Sarah certain rudiments of English syntax and semantics. Sarah strings her plastic words in grammatical order and seems able to answer questions, distinguish use and mention, deal with a range of sentences containing "some" and "all," and handle certain logical operations. Just how far Sarah and Washoe can progress are taken by these theorists to be open questions, limited as much by experimental ingenuity as by the chimpanzees' capacities.

Now all these investigations of animal learning and communication are interesting in and of themselves, but each has been regarded as of wider theoretical significance. Every new instance of animal communication, especially those cases seeming to involve natural language, has been greeted as important evidence disproving the nativist thesis that there are special innate factors critically responsible for man's linguistic competence. This attitude is shared by friends of nativism as much as by its foes, for the nativist response typically has been to deny that what the animal communicates with is really a language, or that the limited symbol systems taught the chimpanzees have anything significant in common with English. But it is not clear why some animal's mastery of a natural language should contradict the theory of either the nativist or the nonnativist. If the antinativist rejected nativism because his learning theory required that any organism capable of learning could learn anything, then he might indeed insist on the existence of talking chimpanzees as important evidence in his favor.[1] But these grounds for rejecting nativism would seem also to imply the existence of talking foxes, goldfish, and amoebae, and this result is not one most nonnativists would be willing to subscribe to. Rejecting nativism ought to be compatible with a learning theory that can admit to differences in the learning capacities of humans and goldfish. Yet it is no more obvious why the existence of talking animals should present a challenge to the nativist's opposition to nonnativism. If organisms other than humans demonstrate an ability to talk, a reasonable nativist conclusion would seem to be that organisms other than humans possess the innate structures that underlie this ability. Surely the nativist has no reason for wanting to maintain that anything innate must of necessity be limited to one and only one species.

Of course, the link between nativism and species specificity becomes apparent when the nativist, as he frequently does, allies his thesis with the more metaphysical claim that possession of natural language is *a* or *the* distinguishing feature of human mentality. For if man's uniqueness among beasts is to be characterized in terms of linguistic capacity, then it is obvious why someone maintaining this position should be bothered by talking animals. What is not so obvious is why a nativist theory of language entails or is entailed by species specificity, and why the nativist, as opposed to any other theorist, should be the one to see natural language as definitive of or essential to human rationality. If the animal studies are to provide empirical grounds for settling the nativism controversy, these presuppositions or implications of nativist theory must be brought to the surface. And to do this, we must first have a clear statement of what sort of psychological theory a nativist account of language development could be.

Language competence could plausibly be linked with the *nature* of the species possessing the competence if, when the nativist claimed that such competence was innate, he meant it was present at birth; for, under the circumstances of so limited an amount of experience, there is little else available to explain the possession of such a competence except the constitution of the species in possession. This traditional expression of nativism opposes innate skills or behavior patterns to those acquired by experience; competences that are innate can be

---

[1] Throughout we shall use the word "talking" to cover any cases of language mastery regardless of whether the medium be sound, movements, plastic signs, etc.

identified because no process of learning is needed. But, of course, this version of the nativist theory achieves its clarity at the expense of its veracity. The claim that language competence can be accounted for in terms of properties of the species because all humans talk from birth and no animals do is false, and so there is no need to look for remarkable animals to refute this claim. Nor is the veracity of the claim enhanced by revising it to say that humans come by their language without help from experience via some process like matuiation. For, if it is false that humans come into the world equipped with a natural language, it is equally plainly false that they will develop one without a rich experience of a linguistic sort. The mental structure of each human cannot be so powerful as to tell the whole story with respect to man's linguistic skill.[2]

The nativist, however, still might want to distinguish what the animal does when taught a language from what the human does. He might argue that human language development is the result of a structure so detailed and highly specific that it is plausible to maintain that, although experience is required, the process involved in language acquisition is not one of learning after all. Rather, the process by which humans come to possess language might be thought to make use of experience in the way that the process of imprinting makes use of inputs from the environment.[3] In the case of imprinting, experience does play a role in shaping what is acquired, but only in the sense of triggering or filling in some preexisting set. For example, some birds are said to be imprinted with the "concept" of a conspecific, rather than learning from experience what sort of species it belongs to. This is because, throughout its life, the bird will apply the behavior responses appropriate to a member of its species to whatever sort of moving thing it first spies after birth. Thus, although the content of the "concept," the actual object to be followed or mated with, is discovered by the bird in his environment, the form or pattern of the "concept" is part of the built-in structure or program of the bird. If language acquisition is to be described after this model, then it must be that we are to assume that humans come into the world equipped with the form of language (some would say the rules of universal grammar), whereas they discover the content, the particular nuances of their native language, in their own peculiar linguistic environment.

---

[2] This fact about language acquisition clearly differentiates it from other skills or habits also claimed to be of innate origin, such as reflexes, certain perceptual constancies, or the pecking behavior of chickens, all of which might reasonably be held to be present from birth or to be a matter of maturation. Although often cited, the relevance of these cases to a discussion of language acquisition, where competence is clearly neither present at birth nor a matter of maturation, is quite minimal.

[3] See Lorenz (1965) or, for a concise statement of the issue, Lorenz's paper, "Imprinting," excerpted in Birney and Teevan (1961). That the imprinting model and other similar models that postulate preformed pattern sets or innate release mechanisms have influenced nativist theories of language can be seen in the frequent references these theorists make to the works of Lorenz and other ethologists.

But can such a model derived from imprinting be shown to square in an illuminating way with the processes we observe to occur when a child acquires language? In the case of the imprinting bird, a relatively clear distinction between form and content can be made to seem applicable. The use of this distinction in the imprinting model is justified by pointing to the early occurrence of a behavior pattern that defines something as a conspecific, and to the inviolability of the behavior pattern once an object has been selected. No matter how inappropriate the object, the pattern of behavior is not corrected. But incorrigible rules of this sort do not appear to be a feature of language learning. Furthermore, this model, or other similar models postulating innate release mechanisms, would seem to require that humans be preset for language in such a way that mere exposure to a few instances of a particular language would be sufficient to trigger the acquisition device. No processes of trial and error, mistake and correction, supplementation and deletion, gradual change and improvement, etc., would be expected. But this view of how man acquires mastery of a language seems only a little more reasonable than postulating the competence full blown at birth. The grammatical rules that characterize adult speech are not observed to crop up immediately in the linguistic behavior of the child. Instead, mastery of even the rudiments of adult semantics and syntax takes a period of several years. During this time, the child is continually bombarded with relevant experience he uses and incorporates with varying degrees of speed and success. He makes a number of false starts, improper substitutions, and misguided supplementations, and in the process of correcting mistakes shows himself to be sensitive to his continuing linguistic environment. There is, in fact, little reason to argue that human language acquisition is more like imprinting than it is like other cases of learning.

Here again, the animal learning experiments are not able to clarify matters. People have, for example, wanted to argue that the slowness with which chimpanzees acquire anything like a language suggests that they require something very different in the way of relevant linguistic experience from what human children use. But even if it could be shown that chimpanzees required a much richer teaching environment than humans in order to learn, this would not in any way entail that humans do not *learn* language, that acquisition is merely a matter of triggering or filling in a preset competence. So long as we take a balanced view of what goes on when language is acquired, it would be false to say that language is innate if this meant either that linguistic competence required no experience or nothing like learning for its acquisition. And this means that if nativism is to be an intelligible model of language development, it must be seen, not as an alternative to learning theory, but rather as part and parcel of some learning theory.[4]

---

[4] In fact, it may well be the case that reflexes, instincts, and imprinting patterns are all shaped by experience, so that the behavior we typically observe is not untouched by processes of learning. Perhaps, then, there are no pure cases of structured or developed behavior patterns that are present at birth or merely require triggering to emerge. But if this is so, it still would not alter our point, for it would just show that not only language, but even simple instinctual behaviors cannot be described as independent of learning. For a discussion of this controversy see Hailman (1969) and various articles in Birney and Teevan (1961).

But how is the nativist to incorporate the claim that language or the form of language is innate while still allowing for the importance of the linguistic environment? One approach has been to argue that the mental structure that provides the form of language makes its presence felt as a restriction on the types of languages that can, in fact, be learned. According to this view, the form of natural language (again, some would say this is specified by universal grammar) is innate in the sense that languages not of this form cannot be learned by humans (or could be learned only with such difficulty and expense of time as to be beyond human grasp for all practical purposes). Systems that violate the given form are unlearnable.

Here it should be tempting to ask the nativist what evidence he has for his claim. Has this limitation hypothesis ever been directly tested? Is it even testable? Certainly there is no evidence presently available that would show that a language just like English but violating even one universal feature could not be learned. Nor is there any established theory of cognition that indicates that humans could not readily master some artificial language consisting of, say, a two-word vocabulary and only one rule of syntax, even if that rule were not of the constrained sort. Indeed, the evidence we do have of man's ability to break codes of serious or fanciful sort, uncover the regularities underlying varied series of numbers, pick up game rules from watching the play, etc., suggests a more charitable view of human cognitive capacities than this version of the innateness claim allows. But perhaps even more to the point than consideration of man's general conceptual abilities, is that, even under quite ordinary circumstances, most humans master a multiplicity of symbol systems seemingly not of "the form of natural language." It is very easy to think of quite a lot of symbol systems such as gestures, maps, diagrams, pictures, music notations, graphs, and imitations that are learnable but fail to answer to the universals that can plausibly be provided for natural languages. Nor are there grounds to suppose that all these systems could not be mastered without the use of a natural language. Indeed, acquisition of some of these systems, like gestural, pictorial, and imitation schemes, usually precedes mastery of natural language and most likely plays a role in its development (Piaget, 1968; Sinclair-de-Zwart, 1969).

Perhaps the nativist might want to counter all this evidence by replying that he never meant the structural restrictions on learnability to apply to other sorts of symbol systems besides natural language. Therefore, that humans can learn artificial languages, make and break codes, and master various nonlinguistic symbol systems cannot refute his claim. But such a reply would leave this version of the nativist claim most unclear. Can it be that only systems of the specified form are to be considered natural languages? If so, then it is obvious that natural languages not of this form would not be learnable, for there would be no such natural languages. Such a trivial claim would be weak grounds for what is supposed to be a new theory of language acquisition. Clearly, the experiences we have that lead to our having acquired a natural language, lead to our learning English rather than some other language that would violate the supposed restrictions of universal grammar. This much is tautologous. But this does not in any way imply that, in other circumstances, with different experience, other symbol systems or languages not of the specified form could not or would not be learned. There seems to be no reason to believe that we will either learn languages of a particular form or nothing at all. That some of these other kinds of systems might be harder to master than natural language or that a language just like English but violating some "linguistic universal" would take longer to learn than English need not be denied. Not all symbol systems, like any other kinds of skill, need be attainable at the same rate. And the point of the limitation claim is further diminished once we admit that natural languages themselves take a considerable amount of time to learn, and it seems reasonable to allow that some other very simple "deviant" system could be learned more quickly than they can.

Even if some version of the limitation interpretation of nativism could be made plausible, moreover, it is unlikely the animal studies would be of much relevance. For to establish that some animal other than man can or cannot master English says nothing about what sorts of systems unlike English humans will be unable to master. Thus, construing the innateness hypothesis as telling us what cannot be learned seems unacceptable on empirical grounds. Such a theory, moreover, really gives us no purchase on an account of how language is acquired. It would only give us a reason why some possible symbol system was not in use— it was one that could not be learned. But if there is thought to be some feature peculiar to the human mind, it is that, with appropriate experience, human beings *will* learn a language. It is this ability to learn the languages that they do learn, then, that, as nativists, we would want to appeal to mental structures to explain.

What the nativist must provide for us is not an account of how mental structures prevent some languages from being learned, but rather an account of how such structures play a role in determining the language that is learned. Now the current cognitivist model of language use and acquisition is seen by some nativists as providing just such a framework in which to develop a distinguishing position for himself. On this account, learning a language is seen as acquiring (mastering, implicitly knowing) a set of hypotheses (the grammar) that generate and characterize the language. Knowledge of these hypotheses is supposedly what enables the learner to speak and understand the language. And the task he has accomplished is said to be one of generating and choosing hypotheses on the basis of the evidence with which he is presented. Indeed, the child learning his native language is most often described as being in a situation analogous to that of a field linguist attempting to construct and write a grammar for an unknown language.

This task of choosing the correct hypotheses, however, is thought to be a very difficult task once we take into account the pervasive effects of ordinary inductive indeterminacy. For, as we know from studies of inductive logic, there will always be countless alternative conflicting hypotheses compatible with any nonexhaustive set of evidence. Therefore, to learn a language, the child must be in a position to choose from among the set of hypotheses all compatible with his evidence and differing in their future projections. It is only reasonable, the nativist argues, that we assume the existence of mental structures determining the child's selection of one set of hypotheses rather than some other conflicting set. After all, what else is left? If the child is not taught the rules explicitly, he must make the inductive leap on his own. And if we consider that all speakers of a language must end up with the same set of hypotheses, even though the evidence (corpus of instances) each has encountered is very different, the only reasonable way to account for this uniformity is on the basis of innate structures determining the selection (Katz, 1966).

There are several formal points to be made concerning the relationship between this model and a nativist theory. The first is obviously that, if a nativist is willing to commit himself to this model, then he has no reason for wanting to restrict its application to human beings. The inductive problem that mental structures are introduced to solve will emerge whether or not animals are capable of fluent English. If the data underdetermine the hypotheses and if learning requires that the learner choose among hypotheses, this task will remain no matter how large or small the set of creatures capable of making such choices turns out to be. For the problem of learning raised by this cognitivist model is purely the formal problem of relating a finite set of data to an unbounded set of conflicting hypotheses. Further, since the underdeterminacy of hypotheses remains whenever the data are inexhaustive, talk about how slowly or fast language learning takes place is also beside the point. Whether it takes 2 years or 10 to master a language, the instances experienced in any case cannot uniquely determine the hypotheses. Even at the end of 10 years, logically there will still be an unbounded set of conflicting grammars from which to choose. It is hard, moreover, to see why the model has any special application to natural language. Clearly, the problem of learning almost any symbol system, when the teaching is by instances, can be described in its terms. Indeed, the scope of this model would seem to extend far beyond symbolic capabilities; in the acquisition of almost any sort of skill the hypothesis needed to generate and characterize the competence eventually possessed far outruns the data available to the learner.

But the scope of the model ultimately depends on the possibility of its realization in empirical terms. For it is important to recognize that, at least so far as language learning goes, the model is thoroughly metaphorical. Hypotheses, at least from the standpoint of inductive logic, can themselves occur only as part of some language or system of representation. An organism cannot actually entertain hypotheses in a stage of its development that necessarily precedes the symbolic. To make literal sense of an inductive logic model dealing in evidence and hypotheses, there would have to be another prior symbol system in which to formulate the evidence and hypotheses. The hypotheses and their evidence cannot be formulated in the language being learned, for it is just this learning event that the hypothesis model is supposed to explain. And, of course, even if we assumed that the evidence and the hypotheses were available in some prior symbol system, the model as an account of how symbolic skills are learned must break down, since it leads to a regress. We will need a prior system for each symbol system in which the hypotheses for learning are encoded, in which to encode the hypotheses generated in the learning of that system.

The analogy between a language learner and a linguist provides a metaphor for language learning that becomes increasingly difficult to interpret as it becomes further extended. If we could find some sort of empirical interpretation of the notion of hypothesis that frees it from its linguistic connotations, it will presumably have little in common with the linguistic-based hypothesis that is a projection from a set of evidence. For if these "hypotheses" are to be literally based on *evidence*, the language learner must be held to be codifying the environmental inputs into evidence statements from which hypotheses are generated. The world does not constitute evidence; only statements, that is to say, conceptualizations of the world, can be evidence for a hypothesis.[5] Application of the inductive model would require seeing the child as classifying, or in some way codifying, the sentences presented him as *instances* of the hypotheses they will be taken to support. Thus, prior to mastering any grammar, the child must assign the structure to particular instances that the grammar as a whole will assign to sentences of this sort. But what interpretation of this model could possibly lend plausibility to the claim that a child, even before he can understand a language, codifies input into noun phrase (NP), verb phrase (VP), object of, etc., that "his" future grammar will assign to these sentences?[6]

In general, it is when we try to talk about how the language learner behaves in the process of learning a language that the evidence-and-hypothesis model lets us down most severely. There can be no literal sense made of the claim that the learner *has* evidence from which hypotheses are generated, nor can it be said that the learner *has* hypotheses. Certainly, he cannot be said to have these conflicting grammatical hypotheses in such a literal sense that he could be said to choose among them. Clearly, if there is no obvious way of dealing with the claim that the child has put himself in possession of even one hypothesis, or recognized evidence in its favor, the suggestion that *he* is choosing among many hypotheses in his possession is quite mystifying.

This is not to say, of course, that there is no similarity between the behavior of a linguist projecting hypotheses from evidence and the behavior of the learner. It certainly is true to say of the language learner that he ends up learning something, the English language, for example, and that this something he learns is both intimately related to the data he has available to him, an English-speaking environment, and is underdetermined by these data. It also seems reasonable to talk about the child's handling and producing new cases in the particular way that he does as generalizing on the basis of an inexhaustive, finite sampling of the language. And, since it is not a priori necessary that he "generalize" in one way rather than some other, that his competence or his behavior take the form that it does, something must determine the direction of the learning process. So, here,

---

[5] The point is that a given object, for example, a diamond, is not evidence for "All diamonds are hard." Evidence for this hypothesis are statements, propositions, beliefs, etc., of the form "Diamond x is hard."

[6] Sidney Morgenbesser has also pointed out that, in using the inductive model to account for the generation of "hypotheses," there is a tendency to confuse questions concerning the context of discovery with the problem of justification. Inductive logic attempts to evaluate hypotheses or tell whether they are supported by the evidence; it does not pretend to tell how we come by the evidence, or account for which hypotheses will be generated, given the evidence.

perhaps the nativist has, at last, come upon firm ground. Learning cannot occur against a true tabula rasa. In order to learn language, we must be predetermined by a particular mental structure so that we generalize or go on to handle future instances one way and not another. This predetermination can result only if our brain has a built-in bias, and it is this critical biasing factor that constitutes the innateness of natural language. To say that the form of natural language is innate means that innate mental structures determine that, when we are given a sufficient sampling of sentences, we project or generalize to grammars of one form rather than alternative conflicting forms equally compatible with the teaching corpus.

Though this interpretation of the innateness thesis may provide the nativist with a clear and empirically tenable position, its significance as an interesting account of language acquisition seems limited. For, once we strip from the notion of a critical biasing factor the surrounding aura of the inductive model, with its "evidence," "hypothesis," "selecting or choosing," "ordering of alternatives by a simplicity measure," or whatever, there is not much that remains of the innate mental structure whose existence we granted. We have already found that we cannot conceive of this innate factor as a *source* of language competence that waits only for birth, maturation, or a few triggering experiences before manifesting itself. Nor is the particular "end product," the particular symbolic competence we achieve, predetermined by the innate factor. For it is not as if we will end up either with a system of the predetermined form or with nothing at all, no matter what experiential input we have been presented. Given some experiences, some sorts of environmental input, a learner will acquire the English language, with different input, French, with still different experiences, some nonnatural language or symbolic system, and, if the input is different enough, with the ability to play chess. All that is predetermined is that, given enough of the right kind of experience, we will wind up with linguistic skill, and some features of the brain must be responsible for enabling us to "go on" in the appropriate way. But, now that we have cashed in the evidence-and-hypothesis model in terms of "what is learned" and "input available," we cannot claim to have devised a model that does much more than pose the problem of how learning itself can occur. That we learn what we learn is undoubtedly due in part to innate factors, but this does not go to show that *what* we learn is innate.[7] And if in acquiring language we do generalize to new instances on the basis of old, we should ask what else anyone could have ever meant by *learning,* but that, when

---

[7]Thus to claim that a language or a particular set of languages, or forms of language are innate, that is, that the subject matter that we learn or come to know is innate, could only be misleading. If anything is innate in this context it is not the end product or what we know, but only those mental features which allow us to learn. And if this is correct, it would be confusing to call these learning capacities, innate knowledge or ideas; for this suggests that the ideas, the subject matter, is somehow predetermined or preexisting. The ability to acquire knowledge or an idea is not itself knowledge or an idea. This may well be Locke's point, since he never intended to deny that the mind has unlearned biases, habits, and propensities.

exposed to a particular environment, an organism who has learned some skill can "generalize," "project," or "is able to carry on" in the right way. If we were the type of organism that could not carry on appropriately, that is, in a grammatically acceptable manner, having been given a suitably rich linguistic environment, we would not have, for example, a linguistic capacity. For the ability to learn almost any skill is nothing more than the ability to go on correctly. The gloss the nativist adds is that there is no a priori connection between our experience and how we carry on or generalize. Logically, at least, we could always have done otherwise, and that we do things the way we do must depend upon the type of brain and nervous system we have. But this is nothing we needed the nativist to tell us.

All this is not to say there is nothing new that the nativist believes he has to tell us, or that there is no position that the nativist, with his talk of mental structures, is trying to take up in opposition to the nonnativist. For, if everyone is willing to agree that learning a natural language depends on genetically given structures, there is less agreement about how task-specific are these inherited features which provide men with the capacity to learn a natural language. The nativist's stress on the importance of mental structures seems to result in his willingness to argue that the requisite structures are quite specific and restricted to natural language acquisition, whereas the nonnativist is more willing to commit himself to the view that inherited structures are more a matter of man's general cognitive capacities. Thus Chomsky at various places argues against identifying the innateness of universal grammar with more wide-ranging intellectual capacities or "general multipurpose learning strategies," and even suggests that there is a faculty specific to the acquisition of natural language, distinguishable from, for example, a mathematical faculty (Chomsky, 1968; "Noam Chomsky and Stuart Hampshire," 1968). Similarly, Miller (1970, p. 183) suggests that the problem of innateness be "redefined around the conjecture that there are innate language-specific mechanisms unique to human beings." The commitment on the part of nativists to mental equipment that is task specific, then, would seem to be an issue of disagreement for which the animal studies might be thought to provide relevant empirical data.

But even at this point it is necessary to be careful in interpreting the significance of the results of attempts to teach natural language to animals. Mere failure of some species to master English would not itself show that man's language ability is to be distinguished from his more general cognitive capacities. The nativist could gain support for his position from the failures of a particular species only if he were to assume that the animals are otherwise as cognitively gifted as humans and that failure to learn is not due to other features, such as lack of interest and motivation, or to the use of teaching methods poorly suited to the species.[8] In fact, it would appear that the separation of linguistic capacity from intelligence, by postulating, for example, a distinct faculty, might be argued for

---

[8]The possibility that failure in many cases is a function of teaching methods is brought out in the works of Premack and Gardner and Gardner.

more strongly if some admittedly stupid animal, like a turkey, could master English. But in order to attribute mastery of a language to some animal, the animal would have to be able to apply the symbols of the language appropriately and act appropriately with respect to the statements in the language. And, of course, to the extent that an animal were to use a language appropriately, that is, in an *intelligent* manner, we would be inclined to raise our view of the animal's intelligence. So it is not very easy to dissociate an assessment of a capacity to learn a language from an assessment of intelligence as a whole.

The nativist's difficulties in maintaining task specificity are further compounded in the case of man, for he must somehow separate man's general ability to deal with complex symbol systems of all sorts from the particular capacity to master natural language. Now one way the nativist might use animal language studies to push for this distinction would be for him to produce an animal unable to master English, but able to master some artificial language that he, the nativist, is willing to admit is of comparable complexity and expressive power. But the nativist, typically, has not come forward with evidence that animals without natural language capacity are able to handle languages as "rich" as natural language but not of the form of natural language. Nor does the nativist bring forth evidence that animals without natural language capacity are well equipped to master various nonlinguistic symbol systems such as maps, graphs, diagrams, mathematical systems, etc. Instead, the nativist is frequently the one who is at pains to point out how limited and how "stimulus bound" are most systems of animal communication (Chomsky, 1968, p. 61).

Then what is it that inclines the nativist to ignore the area in which his interests apparently lie? The fact is that it is not at all clear at this juncture where the nativist's interests ought to be. For suppose he were right in claiming that, in order to account for man's capacity to learn natural languages, we must postulate other highly specific genetic endowments that can reasonably be seen as separate from or independent of man's general intellectual or symbolic capacities.[9] This would amount to maintaining that the additional, genetically inherited features have little to do with man's cognitive capacities to abstract, to see spatiotemporal dependencies, to recognize structural relationships, to organize, reorder, plan, and discriminate. So the more it could be argued that those features responsible for natural language *are different* from features that form man's general ability to symbolize and order his world, the less reasonable would be the claim that the species-specificity of natural language accounts for or provides insight into the nature of the human mind and its intensional life, or into man's rationality and, hence, his choices and responsibilities. A psychologist, of course, inter-

---

[9] This, of course, is a large supposition, since even if these additional capacities evolved under the genetic pressures that developed language in man, it would be highly unlikely that these capacities would remain isolated and untapped for many other cognitive uses. We are not claiming, however, that cognitive capacities and skills are unisolable in any sense, a claim that, for example, data on aphasia and other brain damage suggests is false, but only pointing out some difficulties in the claim that these functions are specific to properties peculiar to natural languages.

ested in language learning must postulate the existence of whatever capacities can be shown necessary for language mastery, whether or not any argument can be given for their independence. But to argue that the features responsible for natural language are so highly task specific that they can be separated from cognitive life in general would be to strip the claim that natural language is innately *species* specific of most of its metaphysical as well as its theoretical and philosophical interest.

Acknowledgments

We wish to thank David Rosenthal and Richard L. Herrnstein for helpful comments.

References

Birney, R., & Teevan, R. (Eds.). *Instinct.* New York: Van Nostrond, 1961, pp. 52-64.

Chomsky, N. *Language and mind.* New York: Harcourt. Brace & World, 1968, Chap. 3.

Gardner, B. T., & Gardner, R. A. Two-way communication with an infant chimpanzee. In A. M. Schrier & F. Stollnitz (Eds.), *Behavior of nonhuman primates* (Vol. 4). New York: Academic Press, 1971, pp. 117-181.

Hailman, J. How instincts are learned. *Scientific American,* December 1969, *221,* 98-106.

Hayes, K. J., & Nissen, C. H. Higher mental functions of a home-raised chimpanzee. In A. M. Schrier & F. Stollnitz (Eds.), *Behavior of nonhuman primates* (Vol. 4). New York: Academic Press, 1971, pp. 106-110.

Katz, J. *The philosophy of language.* New York: Harper & Row, 1966, pp. 261-266.

Lorenz, K. *Evolution and modification of behavior.* Chicago: University Press, 1965.

Miller, G. Four philosophical problems of psycholinguistics. *Philosophy of Science,* June 1970, *37,* 183-199.

Noam Chomsky and Stuart Hampshire discuss the study of language. *The Listener,* May 30, 1968, 687-691.

Piaget, J. Language and thought from the genetic point of view. In A. Tenzer (Tr.) and D. Elkind (Ed.), *Six psychological studies.* New York: Random House, 1968. pp. 88-99.

Premack, D. On the assessment of language competence in the chimpanzee. In A. M. Schrier & F. Stollnitz (Eds.), *Behavior of nonhuman primates* (Vol. 4). New York, Academic Press; 1971, pp. 185-228.

Sinclair-de-Zwart, H. Developmental psycholinguistics. In D. Elkind & J. Flavell (Eds.), *Studies in cognitive development.* New York: Oxford Unversity Press, 1969. pp. 315-336.

CHAPTER 9
Talk to the Animals

Margaret Atherton and Robert Schwartz

Evaluations of ape language projects rest on acceptable answers to two relatively distinct questions. First, it is necessary to determine what the studies actually establish about ape capacities. Just what skills and competencies does the evidence make it reasonable to attribute to the apes? Second, and conceptually prior, it is necessary to determine the significance of the accepted findings. Supposing all agree that apes have or lack some particular competence, why should that finding be of interest; to what issue would it be relevant?

The recent proliferation of studies on language acquisition in apes has produced an equal proliferation of criticism (e.g., Gardner, 1980; Sebeok & Umiker-Sebeok, 1979, 1980; Terrace, Petitto, Sanders, & Bever, 1979). Most of this criticism has focused on the first issue, about what in fact the studies prove the apes have learned to do. Critics have argued that the data, when properly analyzed, do not show that the apes have linguistic skill, that there are alternative and more plausible accounts of the apes' activities that do not require attributing to them the mastery of syntactic rules, and that the apes' behavior merely exhibits simple problem-solving capacities. Some have even maintained that the whole set of studies are instances of nothing more than Clever Hans phenomena. Most ape workers, of course, have argued that their evidence supports a much richer evaluation of the apes' accomplishments. Thus the major source of controversy has been over conflicting interpretations of what the data show apes can actually do.

It seems to us, however, that in all this debate not enough attention has been paid to the second evaluation question, namely, the significance or implications of determining ape linguistic capacities. Both ape supporters and critics assume that were it established that apes can master a language this would be a remarkable finding. The reasoning on both sides seems to be something like this: Language competence is intimately tied to, or maybe even definitive of, our concept of human mentality. As long as it is thought that humans are unique in having linguistic competence, there will be grounds for believing that there is something unique and special about human mentality. Perhaps we are the only species it makes sense to claim have "minds." If there is this difference in linguistic capacity between humans and other species, it must be a matter of genetic endowment. Therefore, proof that apes can master a language would hit at claims about the uniqueness of human mentality and its supposed underlying genetic base. On the other hand, the failure of apes to acquire language would weigh heavily in favor of claims that language is innate and would leave standing the claim that human mentality is qualitatively different from that of other species. Ape language projects are therefore thought to be crucial tests of hypotheses about the genetic base of language acquisition and the distinctiveness of human mentality.

It is undeniable that were apes to reach human levels of linguistic competence it would be an impressive feat, and would undoubtedly make us reassess our views about ape intelligence. Nevertheless, we do not attach the same theoretical and psychological significance to the ape studies that both their proponents and critics often do. In particular, we believe the findings can have little bearing on debates over human acquisition and that the typical stress on syntactic skills does not get at what is most important about human mentality.

Language Acquisition and Mentality

Consider the question of acquisition models and innateness. It is trivially true that if we or any other species can learn a natural language it will depend on our genetic make-up. The mere claim that human language competence is a function of innate genetic endowments can not be a matter of serious debate. In Chapter 8 we attempted to explicate various richer construals of the innateness hypothesis and to evaluate the evidence with regard to each.[1] Our conclusion was that when the innateness hypothesis is made reasonably clear and interestingly strong (e.g., a claim about highly *task-specific* constraints restricting or determining what languages are learned) there was not a good deal of evidence to support it. It also turns out, however, that on any of the richer formulations of the innateness hypothesis, the findings from the ape studies are of limited relevance. Although the existence of "talking" apes would certainly decide whether linguistic competence was unique in humans, it would have little direct bearing on the innateness hypothesis. Uniqueness and innateness are separate issues. The innateness hypothesis is a hypothesis about human acquisition that might be true whether or not another species can acquire a language.

Suppose it were shown that apes could master an impressive amount of English.[2] If the course of their language development were different from that typical of human acquisition, nothing much would follow about the factors responsible for our linguistic capacity. Alternatively, were the course of ape acquisition remarkably similar to the ways humans learn language, the obvious move of

---

[1] Our remarks here about acquisition and innateness rely heavily on the analysis and arguments developed in Chapter 8.

innateness hypothesis defenders should be to extend their claim to include apes. For the arguments given for the innateness hypothesis are driven by the claim that it provides the best explanation for the sort of acquisition patterns humans display. To the extent that these arguments are sound, all that ought to be required is to extend their application. Again, innateness and uniqueness are separate issues.

If ape linguistic success will not serve to disprove the innateness hypothesis, neither will ape failure directly support it. The apes may not learn a language for reasons that have little to do with "essential" features of their genetic cognitive endowments. They may not be motivated to learn languages, or, as attempts to get apes to learn *spoken* languages showed, they may lack capacities that are not deemed of central importance. However, even if these "less cognitive" obstacles to acquisition can be eliminated, ape failure would not imply much about what accounts for human success. It would not show that humans succeed in learning languages because they come equipped with highly task-specific structures, such as wired-in knowledge of grammatical universals. It is equally plausible that humans might succeed where apes fail because of more general cognitive superiority. Humans may have better overall abilities to abstract, to analyze underlying rules, patterns, and dependencies not based on perceptual likeness, to see and make inferences, to store symbolized information in memory, etc. In addition to these quite general capacities, humans may also be endowed with certain other cognitive structures and propensities more specifically related to symbol acquisition and use. However, these capacities may not be of the highly task-specific type postulated by the innateness hypothesis. They may be skills or knowledge bases that underlie our ability to master complex symbol systems of a wide variety of types, including systems not readily characterized by the supposed principles of universal grammar. Unless there are grounds for ruling out these other possibilities, the failure of apes to learn a natural language will not discriminate between innateness hypothesis accounts of acquisition and various of its competitors.

Some of these competing accounts might be eliminated if apes were shown to be able to master and use symbol systems as rich and complex as, say, English, but not having the form or principles of organization peculiar to natural language. Such findings would suggest the need for task-specific mechanisms over and above those capacities usually associated with more general aspects of intelligence and symbol use. As far as we can tell, however, none of the recent studies, anymore than earlier ones, serves to establish an asymmetry of this sort of ape linguistic competence (whatever else they may convince one about the level of

---

[2] Although we make reference to English throughout, our remarks are obviously meant to apply to other natural languages such as French, Chinese, and Swahili as well. To what extent other systems, such as ASL, have the same types of structures and expressive powers as ordinary natural languages is a debated issue. While we offer no analysis of ASL in particular, exploring questions about how differences of form and content of symbol systems relate to matters of mind and intelligence is the focus of much that follows in this chapter.

ape symbolic capacity). At the same time, there is plenty of reason to believe that humans can master codes, artificial languages, maps, diagrams, and other complex symbol systems that are not natural languages. Alternatively, it might be more plausible to suppose that linguistic capacity was distinct from other cognitive capacities if we could find some animal less intellectually gifted than an ape that could master a natural language where the ape failed. As we mentioned in Chapter 8, however, and want to spell out more fully here, it requires some care to make good sense of this possibility.

It is, after all, the link that is perceived between the presence of linguistic skills and the presence of a remarkable degree or kind of intelligence that has made it interesting to ask whether any creature other than human beings can learn to use languages. There are undoubtedly any number of other skills that only humans have mastered but whose possible species uniqueness is of little concern. No one would get excited if it were shown, for example, that an ape could mix a dry martini, even if this skill had hitherto been limited to the human species. Language competence has been singled out for special attention because it seems so closely tied to thought and to other distinctive features of human mentality. It is not necessary to identify thought with its linguistic expression in order to appreciate the crucial role symbols play in enabling us to think about the distant past, the unobservably small or far off, the complex, the theoretical or abstract, and the uncharted future. Language significantly enhances our ability to deal with states of affairs that are neither represented nor reflected in the present stimuli, enlarging our capacity to think about situations that may have no direct bearing on our everyday needs or desires.

Moreover, language behavior itself has often been cited as a paradigm case of an activity that implicates mind. In its ordinary use, language production is not dictated by the environment, talk is neither limited to the present situation nor necessarily about present emotional states. Most of the sentences produced do not in any obvious way involve imitating or parroting previously heard sentences. Furthermore, it is not possible to identify sentence comprehension with specific behavioral responses. The connection between understanding a spoken sentence and ensuing behavior is remarkably loose. For all these reasons, then, having a language is often thought to be of special importance to having a mind. Descartes, for example, believed that this type of linguistic activity showed that human behavior could not be accounted for by mechanical means and must be understood as mediated by thought. Descartes (1649/1970) cited the lack of comparable symbolic activity in animals as one of his grounds for claiming they were mindless. More recently, Chomsky put the Cartesian point as follows:

> If by experiment we convince ourselves that another organism gives evidence of the normal creative use of language we must suppose that it, like us, has a mind and that what it does lies beyond the bounds of mechanical explanation, outside the framework of the stimulus-response psychology of the time, which in relevant essentials is not significantly different from that of today . . . (Chomsky, 1972, p. 11)

Thus, the importance assigned to studies of animal communication has been thought to lie in what they show us about the possible presence of mentality and of an intelligence that cannot be explained by simple mechanical means.

Syntax and Mentality

It is understandable, then, that language has traditionally been singled out from other skills, such as martini mixing or bicycle riding, for special attention. It is also understandable that the claim that apes can master a natural language would arouse excitement. However, unlike most earlier discussions of animal language ability, recent controversies over ape language have put a major stress on syntax. Much of the debate has been about whether the evidence at hand really shows apes have mastered a syntax with the same or similar properties as those of natural languages. Both ape supporters and critics agree that ape studies would lose a lot of their significance if evidence for syntactic skill were not forthcoming. For example, Gardner (1981, p. 43) summarized the argument in this way: "Everybody agrees that apes and humans are biologically different. The *significant* question is whether the difference is so large that apes cannot grasp simple syntax" (emphasis added). The idea seems to be that unless apes exhibit syntactic competence, their activity will not be truly linguistic, and hence will not implicate mind. The latter claim, however, runs together issues that are best kept separate. Even though mastery of syntax is part of what is usually meant by knowing a natural language, it is not at all obvious that either the success or failure of apes to exhibit syntactic patterning will serve to establish strong theses about ape mentality. To see why this is so requires some general remarks about the relationship among syntax, language, and mind.

We have seen that language is important for mind in that it enhances the richness and complexity of our thinking, thus helping to make our behavior more flexible and freeing us from direct environmental control. In turn, it is these aspects of human behavior that serve to argue against simple mechanical accounts of our activity and that are taken to implicate thought and mind. If the important links between language and mind lie here, however, evidence concerning the presence of syntax is neither sufficient nor necessary for attributing these features.

It should be quite clear that by itself syntax cannot ensure richness or complexity of thought. Unless the speaker possesses a reasonably developed or sophisticated vocabulary, the expressive power of the language will remain limited. A language, for example, could contain English syntactic rules, yet have a vocabulary that only reflects simple and immediate surface features of the environment (e.g., observed color and relative size) or simple present needs (e.g., hunger or thirst). The sentences of such a language would not sustain complicated or abstract thought. In addition, the presence of syntactic regularities in verbal activity need not indicate that the behavior is either creative or stimulus free. Behavior that displays syntactic patterning could still be readily explicable in terms of mechanical triggering devices. Suppose we find that the honey bee dances exhibit rules of well formedness. Suppose too that the interpretation of one bee's dance by another bee depends upon a fixed sequence of movement in the dance, such that distance must come first, direction second, and degree of sweetness third. It seems unlikely that this appearance of syntax would appreciably affect our willingness to attribute a nonmechanical, stimulus-free mind to the bees. Analogously, we might be able to program a computer to respond to some circumscribed set of inputs in an entirely fixed way, and yet have the computer only respond both to and with sentences that obey the rules of English syntax. The appearance of syntactic regularities, therefore, need not indicate the presence of interesting thought, nor point toward flexible, nonstereotyped, thoughtlike mediators of behavior.

It is also unlikely that syntax is *necessary* in order to secure the link between symbol use and mind. Maps, models, graphs, pictures, musical notation, and diagrams, to name but a few, are symbol systems whose expressive power is not limited to describing surface features of the here and now. Mastery of these systems enables us to use these symbols creatively to represent an unlimited set of novel situations and to understand new symbols or representations never previously encountered. Ordinary human use of these systems is not stimulus bound. There is, in general, no fixed relationship between using or understanding these symbols and present stimuli or ensuing behavior. Thus reasonably rich and interesting ideas may be freely expressed without employing standard sorts of syntactic devices.[3]

This is not to deny that syntactic languages, such as English, may have greater expressive potential or allow for some "thoughts" not readily put in these other systems. It is just that the connection between symbolic competency and mind does not hinge as crucially on syntax as debates about animal communication might lead one to believe. Furthermore, it is not at all obvious how many of the elaborate structures and principles found in natural languages are essential to its expressive power. Richness of thought is more a matter of content than of form, while many of the syntactic rules of natural language are more concerned with intricate ways of embedding and varying basic ideas. We can leave as an open question how much of the essential expressive power of, say, everyday English, would be sacrificed if there were no passives, nominalizations, or deletions, if various clause embeddings and other recursive devices were eliminated in favor of simple sentences and perhaps their Boolean compounds, if certain grammatical relations were marked semantically rather than relying on features of construction, etc. In any case, a good deal could be said in a system lacking

---

[3]Much is often made of the infinite productivity of the syntactic rules of natural language, but this feature is not as special or as significant as it may seem. Other symbol systems allow for unbounded productivity too. While principles of unlimited recursion are important when considering the logical and mathematical properties of natural languages, placing finite limits on such productivity would take away little from our ordinary use. For more on this issue see Schwartz, 1978.

these features, a system that might, then, be characterized by simpler types of grammars than those typical of natural languages. Therefore, it would not be necessary to show that animals can master natural language syntax in order to establish that their possible symbolic activity allows for the creative, stimulus-free expression of thought, and hence implicates mind. (For more on the problems that arise from claims linking creativity and mental processes to syntactic competence, see Drach, 1981.)

Of course, one thing that is impressive about human syntactic accomplishments is that we can master and use a very intricate syntax involving complex structures, relations, and dependencies. To the degree that ape linguistic activity does not provide evidence for the rich sorts of devices found in natural languages, we might refuse to admit that they really have natural language competence. Such failure would also lend support to the claim that humans are unique in having a language that requires sophisticated syntactic capacity. In turn, the human ability to master and use elaborate syntactic devices might itself be seen as an indication of intelligence. It would seem plausible to suppose that being able to use, as we do, a system with the complexity of English would require a certain richness of analytical powers, memory structures, and other cognitive skills. Furthermore, the fact that we can *acquire* this complicated syntactic competence on the basis of limited data, that acquisition is not the result of direct imitation, and that we can develop the ability to produce and understand new structures never previously encountered, would seem to attest to impressive mental power. Nevertheless, learning the syntax of natural language is not the only difficult intellectual task humans can master; thus there is little basis for seeing particular syntactic accomplishments as *uniquely* symptomatic of the richness of human mentality.

When proponents of the innateness hypothesis, moreover, argue in favor of the thesis that language acquisition is a genetically based endowment of the human species and indicative of the structure of the human mind, it is for reasons that tend to undermine the claim that learning syntax is an intellectually impressive accomplishment. In innateness hypothesis accounts, language learning is said to require much that is preprogrammed or genetically determined. Acquiring syntax from the impoverished data standardly available is claimed to be such a difficult task that no degree of intelligence, general problem-solving capacity, or analytical ability—at least not that possessed by an ordinary human—would be enough to account for our success. However, the more mastering syntax depends on wired-in, highly task-specific structures, the less demanding it would be on our cognitive ingenuity. To the extent that acquisition is understood on the model of "triggering preset structures" or "resonance" or "like growing an eye," the more "mechanistic" syntactic development itself turns out to be. The acquisition of syntax would not require and thus could not serve as evidence that the human mind is flexible or particularly creative or inventive. Consequently, there would be fewer reasons to assume any interesting link between our mastering natural language syntax and having a mind. For the acquisition of syntax under the innateness hypothesis is not an indication of mind at its finest, but rather of a mind that might make a mechanist happy.

Conclusions

If the links between language and mind are conceived along the lines we have argued for, perhaps we can begin to appreciate why recent ape communication studies have resulted in extended inconclusive debating. It is difficult to see how the sorts of studies run could be interpreted as "crucial" experiments about the nature or level of ape intelligence. Neither ape success or failure at these tasks can provide anything like conclusive evidence for grand metaphysical claims about whether apes have nonmechanistic "minds" or whether human mentality is "unique."

It may well be the case, for example, that the critics are right and that the language abilities demonstrated in the ape studies can be accounted for by relatively straightforward means, as instances of conditioned responses or as the result of cuing. This would by no means prove, however, that ape behavior in general can be accounted for by simple stimulus-response mechanisms or without postulating more elaborate cognitive mediating states. Even if the linguistic tasks demanded of the apes in experimental settings do not require very much in the way of intelligence, the life of the ape in natural settings may ask for and show a good deal more. It is possible, as well, that apes may use (or be able to use) various rudimentary systems of representation that can enhance their ability to deal in more flexible stimulus-free ways with their environment. These systems, however, may lack the syntactic and semantic features researchers have been attempting to teach the apes.

We also believe that far too much attention has been devoted to debating issues such as the following: "Is what the apes exhibit really a language?" "Does it contain words?" "Does it have syntax?" As Chomsky recently pointed out, the question of whether to classify a system as a language is not so much an issue of empirical science as it is one of adopting a convenient convention about how the word "language" is to be used (Chomsky, 1980). (Chomsky also argued that studies of ape language learning are unlikely to shed much light on human acquisition.) It is not a settled matter, for example, which of the wide variety of symbol systems humans use are to be considered languages, and why; nor is it clear what would hang on such decisions. Plainly, there is a range of signal, symbol, and communication systems, varying in form, structure, and semantic devices, and differing in expressive power and complexity along a variety of dimensions. However, as we have argued earlier, judgments about mentality will depend primarily on the richness and complexity of the messages conveyed and the use to which linguistic skill is put. If, as in most of the current studies, ape conversations are limited to rather simple labeling and demand tasks, these communication episodes will not be indicative of a rich mentality. This will remain so whether or not we decide that the conversation consisted of "words," arranged in "syntactically grammatical linguistic strings."

Finally, it is most important to keep in mind that it is not just spouting sentences in a creative stimulus-free manner that reflects thought. Human language use suggests the presence of mentality to the extent that humans provide evidence that they *understand* what they are talking about. Language comprehension, however, is not an all-or-none phenomenon; it consists of and is shown in a multitude of interrelated linguistic abilities. A developed understanding depends upon such skills as knowing what inferences to draw, knowing what evidence is relevant to more abstract and observationally removed propositions, being able to participate in certain social practices of reason giving and justification, and being able to use language to help plan and guide activity. Language competence, like intelligence and mentality, is therefore best thought of as a graded notion, expressed in a range of interlocking cognitive skills, that may be developed to varying levels of sophistication. Thus it is unreasonable to suppose that isolated studies of ape skills on a few linguistic tasks could possibly tell us anything definitive about the presence of "language" and "mind."

References

Chomsky, N. *Language and mind* (Enlarged ed.). New York: Harcourt Brace Jovanovich, 1972.

Chomsky, N. Human language and other semiotic systems. In T. A. Sebeok & J. Umiker-Sebeok (Eds.), *Speaking of apes*. New York: Plenum Press, 1980.

Descartes, R. Letter to Henry More of February 5, 1649. In *Descartes: Philosophical Letters* (A. Kenny, Ed. and Trans.). London: Oxford University Press, 1970, pp. 243-245.

Drach, M. The creative aspect of Chomsky's use of the notion of creativity. *The Philosophical Review,* 1981, *90,* 44-65.

Gardner, M. Monkey business. *The New York Review of Books,* March 20, 1980, 3-6.

Gardner, M. More on ape talk. *The New York Review of Books,* April 2, 1981, *28,* 43.

Schwartz, R. Infinite Sets, unbounded competences and models of mind. In C. W. Savage (Ed.), *Perception and cognition*. Minneapolis: University of Minnesota Press, 1978.

Sebeok, T. A., & Umiker-Sebeok, J. Performing animals: Secrets of the trade. *Psychology Today,* November 1979, *13,* 78-91.

Sebeok, T. A., & Umiker-Sebeok, J. (Eds.). *Speaking of apes*. New York: Plenum Press, 1980.

Terrace, H. S., Petitto, L. A., Sanders, R. J., & Bever, T. G. Can an ape create a sentence? *Science,* 1979, *206,* 891-902.

CHAPTER 10

Apes Who Sign and Critics Who Don't

William C. Stokoe

More than a decade ago Gardner and Gardner set out to see just how much language behavior an infant chimpanzee might acquire, compared with that acquired by a human infant, under conditions that would be comparable. They provided, therefore, a responsive, human environment, surrounding their subject with human companions who signed instead of spoke. Their chimpanzee, Washoe, is now justly famous, and since then several other nonhuman primates have been taught or experimentally exposed to human signing. A common factor in all these experiments is a theoretical base in psychology, but the similarity ends there. Gardner and Gardner are comparative psychologists and so state. Other experimenters express or reveal adherence to behaviorist, innatist, or eclectic theories of language and mind.

It is my contention that none of these systems (any more than linguistics or psycholinguistics unaided) can really test such an experimental design. To compare human behavior with that of other animals requires first a fully realized human science, something not yet in clear prospect. Only a science of man and woman will be able to test the premise that is often taken as axiomatic in other behavioral sciences—ultimately a theological premise—that human creation was unique and that human behavior, especially language, is cut off from any evolutionary continuum.

Among the sciences now available only semiotics seems to have the detachment needed to offer a venue for the case of language in apes. Semiotics as envisioned by Peirce is the science of signs. *Signs* to semioticians are not the simplistic pairing of something with its meaning but instead have a triple structure: sign vehicles (in the realm of perceptibility) are related to sign denotata (in the realm of cognition) by and through sign interpretants (functioning organisms). Nothing in the foundation of this theory specifies that an interpretant need be human. In fact, much recent semiotic work, notably that by Sebeok, is in zoosemiotics, focused on animal communication.

There is more than the avoidance of anthropocentrism in semiotics to recommend it as a context for viewing ape language experiments. Eco (1976) carefully distinguished between *signification* (conveying information by, but not intrinsically in, signs) and *communication* (mutual exchange of information by organisms). The latter presupposes the full operation of the former, and it is possible that the former may be easier to demonstrate conclusively in nonhuman animals than communication through language. The work of Premack clearly shows that the chimpanzee Sarah operated with plastic tokens as signs. One recalls also films showing Washoe playing alone and making manual signs, recognizable as signifying (to signers of ASL and to others) the names of the objects she is playing with. Communication by use of such signs between apes or apes and humans may be harder to establish, but there is much evidence that it does occur.

Specific Differences in Communication

Signification and communication, then, function at a much more elemental level than language. All living organisms get signs from their animate and inanimate environment. That the interpretant takes information from something perceptible in inanimate nature is sufficient to make that thing a sign, although its sign function is supplied entirely by the receiver. All living organisms also emit what creatures of the same species or others interpret as signs: that is, information to be acted upon. The whole scope of zoosemiotics shows conspecific communicating in a bewildering variety of ways. Different species communicate differently; and it is not surprising that the more physical similarities between a species and the human species, the more similarity in communication patterns appears.

As general observers we have seen little in the reptilian order that resembles our communication, but in the related order of birds, much notice has been taken. Birds communicate with their young. The cheeping and open mouths of infant birds are signs and so is the food placed in their mouths by parent birds, the sheltering wing of an adult bird, the danger call, and so on.

Mammals, however, including the human species, communicate in ways that birds cannot:

(1) Prenatally: the embryo is in direct physical communication with one parent's body from fertilization to birth.
(2) Organically: feeding is no longer mouth to mouth but from a special (mammalian) organ to infant mouth.
(3) Directly: the tongue, muzzle, and head are used to dry, groom, caress, support, lift, direct, warn, and punish.

Every species is sufficiently organized (negentropic) to survive, and even extinct species have left evidence that they once filled an ecological niche. It seems evident too that there is a progression in signification and communication from avian to mammalian manners of communicating. The mammals show an increase of complexity in physical structure and an increase of specialized organs and functions for communication. This apparent progression seems even to accelerate when we consider primates. Primate mothers have the following: (1) arms and hands that are used for carrying, holding, grooming, fondling, and punishing the infant; (2) faces, frontally faced eyes, and the bipedal postural capability that allow visual face-to-face communication *while* the tactile communication proceeds; and (3) voices that can be used *while* tactile and visual communication proceed. (Primate fathers have most of these structural features, too, but primate social organization usually puts a distance between father and infant.) This triple-channel communication through touch, sight, and hearing is of course more complex in organization, but the basic reason that nonprimate communication is unlike it is that nonprimate anatomy does not permit such sign production.

Homo sapiens is a primate too, and human communication adds little physically to the triple channel. If Lieberman (1975) is correct, the chief difference is in the throat: no other primate, not even the human neonate, has a two-chamber, variable-ratio, supralaryngeal vocal tract capable of making certain vowel sounds contrast over the whole range of human voices. The other slight physical difference, however, may be equally or more important—not for speech but for enabling the hominid branch of the higher apes to develop language. This difference at first was one of degree. Most primates can stand and move on two legs, and so may be said to have hands and arms; but man is the only primate to be fully bipedal. Fossil finds now are sorted quite accurately into man or ape-man by differences in lower leg structure that come from regular instead of occasional bipedalism. This regular bipedalism of the human line would of course enhance all those tactile, facial, visible parts of the uniquely primatoid communication, and with this enhancement, the accompanying vocally produced signs in a richer, denser context would become more potent as well.

Although the popular view and many scientific views of human communication focus on vocal behavior, the foregoing should make it clear that in the normal development of the human infant the mother-infant communication is still tactile, kinetic, and visual as well as vocal. (For a description of the extraordinary richness in parent-infant communication in seven deaf families, see Maestas-y Moores, 1980.)

When we turn from communication to language, it is clear that the use of language is still communication. Language itself is the outside aspect of this form of communication; it also has a psycholinguistic inside—the linguistic competence of the individual. The exact nature of the inside is debated: few now would say that the human infant is born a clean slate; obviously much of the history of the species, the genus, and the order is in the genetic inheritance. However, there has been of late some attempt to minimize the effect of experience and communicating in developing linguistic competence, which obviously differs at age 10 from what it was at age 1 in the same individual.

Thus the outside of a language is information—sensible behavior observed by more senses than one—accessible to a detached observer. The inside is information also—competence or the grammar of the language—but deeply inside the organism and not directly observable. Having both inside and outside, language cannot be fully comprehended by either a behaviorist or a mentalist theory. Language partakes of both. During the decade of chimpanzee language research and for a decade before that, research into the organization of human sign languages has made it clear that the inside (human language competence) can be as completely and grammatically transformed into outside behavior when the medium is visible kinetic behavior as when it is audible vocal behavior (Stokoe, 1974, 1978).

Training or Learning?

Sebeok and Umiker-Sebeok (1980) reviewed very thoroughly the whole subject of communication in language with other animals, especially chimpanzees. After their exhaustive study, which makes reference to 175 works, there may be little to add, but that little is still worth adding. First, making use of their clear distinction (from Hediger) between apprentissage and dressage in animal training, I would like to consider my own interactions with two signing chimpanzees. Second, I will try to confront the issue that they say they have intentionally skirted (Sebeok & Umiker-Sebeok, 1980, p. 53): "the consequential issue so competently examined by a number of authors such as Bronowski and Bellugi, Brown, Chomsky, Lenneberg, Limber, and McNeill ... whether what is being taught the apes is really 'language'." Finally, I have a minority view to express: I believe that the good that the ape experiments have done and may do should get a higher valuation than is now current as it would if we were to learn the proper lessons from the whole story.

It is proper for psychologists to raise questions about methods used in the experiments with signing chimpanzees. It is worth reiterating that the majority of the experimenters are psychologists and to note also that each group has severely or savagely criticized all the others. The methods that have been used are all methods of experimental or comparative psychology, but the experimenters may not have considered a much older methodology—that of animal trainers. The discipline of animal training goes back to very ancient times and has been very carefully studied (Sebeok & Umiker-Sebeok, 1980).

Depending on the outcome desired, animal training can be either apprentissage or dressage: the outcome of *apprentissage* is that the animal is taught and then performs—without coaxing, prompting, direction, signaling, or coaching apparent from the trainer—a sequence of difficult actions. The outcome of *dressage* is that the animal performs—with constant, direct, control (often through physical contact) from the trainer—on command, any one of various difficult actions.

Under apprentissage or dressage, animals can do things that astonish people who know nothing about the limits of training. Anyone who wants to teach an animal gestural or other kinds of signs must first know all about apprentissage and dressage. Otherwise, the experimenters will never know whether the animals are doing language or doing their "tricks."

There is also another dimension to animal-human communication: fraud. About 80 years ago a horse called Clever Hans seemed to count and do math by stamping one forefoot the right number of times. Pfungst found that the horse was actually reading nonverbal communication. Observers inadvertantly signaled expectancy and satisfaction, indicating when the horse should start and stop stamping to give the correct answer. The question is, how did they signal?

Pfungst could explain it almost completely by head movements (Rosenthal, 1965; Sebeok, 1979). Pfungst had photographic proof that Clever Hans could detect a head movement as small as a few millimeters. More recently, Timaeus (1973) studied the evidence and says that the horse watched eight channels:

(1) Eyes: blinks, direction change
(2) Head: movement up, down, or side
(3) Mouth: lip changes
(4) Body: changes in postural tension
(5) Hands: movements
(6) Jaw: changes in muscle tension
(7) Voiceless counting along
(8) Breathing: patterns of inhalation.

The experts all looked at the horse to understand how he did it. As Sebeok (1979) said, they should have looked at the human spectators. Sebeok applied this warning to the recent experiments also: everybody examines the chimps to see how they get so smart; they should look at the human element and see what the apes' companions, teachers, and trainers are really doing. Sebeok noted that each group of experimenters criticizes the rest for falling into the Clever Hans trap, but judged all of them guilty.

He may be right, but I want to consider the Clever Hans phenomenon as having two aspects: it can be used to deceive, and it is often used that way. Consequently, most people who pay mind readers, psychics, and mediums are willing victims of fraud, and there is a worse and more expensive negative point: the experimenters may be deceiving themselves. Sebeok is sure that they are. They think their chimps are thinking, not performing apprentissage or dressage; although they may tell us that the other groups' chimps are carefully trained, or are acting like Clever Hans. Scientists who deceive themselves are a danger to science, I agree, but let's look at the positive side of what Clever Hans can teach us.

Extralinguistic Communication

Hans learned a lot about human behavior, and Pfungst and later observers learned a lot about communication between animals and man. New experiments need to be designed to study precisely what communication—not language, but communication in the rigorous definition of signification and communication by Eco (1976)—occurs between us and our ape cousins. I am convinced that the Clever Hans phenomenon need not destroy all the value of experiments with chimpanzees and sign language, but the new experiments must, as Sebeok insisted, take the best available knowledge about animal training into account. It is imperative that we see just how the clever apes are getting information from human sources. At the same time, the experimenters must learn to stop looking for evidence of human thought and human language in the animals and look instead for all types and channels of transmission carrying information that we can learn to read as well as the animals are reading it.

Terrace (1979) made it very clear that some of Nim's 60 teachers could read information from Nim, but gave no explanation at all of the semiotic signs they read or what channel the information came through. He said of Laura Petitto, Nim's best teacher: "Laura dominated Nim by generally being one or two steps ahead of him" (p. 122). He also quoted Laura's diary of her work, about her evaluation of her own attitudes: "It was more than just being relaxed. Nim had an uncanny ability to read one's feelings. I always felt that I had to be honest because he understood me . . . he made me feel 'naked' " (p. 111).

This is exactly the matter that needs to be studied: How do animals understand us, and how can we understand animals? The importance of using chimpanzees as the animals is their great similarity to us. There are good ethologists who study the communication inside a species and across species, even some like Lorenz, who studies communication when a human being is part of the ring. This is a promising way to study the nonhuman primates, because only they have the anatomy that encodes important messages in the channels used by human signed languages. My opinion is less negative than Sebeok's. I think the Clever Hans phenomenon and all that Sebeok knows about animal communication might be put together into some very useful studies of cross-species communication that would serve a broad science of humans in nature.

It is true that the Clever Hans phenomenon has led gullible persons to accept telepathy and other kinds of extrasensory perception very uncritically; it is true that many if not all cases of uncanny communication come back to the Clever Hans phenomenon. However, the kind of reading, understanding, and reacting that Hans actually did is not telepathy; it is sending and receiving messages. Where Sebeok and I disagree is somewhere between zoosemiotics and anthroposemiotics, very close to language itself. I think languages or anthroposemiotic behavior came from zoosemiotic behavior and that the harder we look the more likely we will be to discover how. That would have great advantages. Others believe that the gap is too great ever to have been bridged, and that once across, man could never again look back and would be mistaken to try. I believe that I do not worry unduly about the animal in me, nor fear that what little humanity I possess will be lost if I try to understand other animals better.

Communication at First Hand

I think that Sebeok and Umiker-Sebeok have an execllent idea, but I doubt that apprentissage and dressage explain everything in the best of these experiments. I have a little first-hand experience that I would like to use to explain my doubt.

When I had a few minutes with Moja (then a 2½-year-old female chimpanzee being taught by Gardner and Gardner and their associates in Reno), I felt that the first part of her behavior might have come from apprentissage. I had been told Moja liked a game of "play dog." It worked like this: if she patted my thigh, as if she were making a sign for *dog* on me instead of herself, I was supposed to make like a dog with barks and woof-woofs. Which I did. She then pretended to be terrified, leapt into my arms, and enjoyed a reassuring hug and cuddle. We did the routine two or three times. You could say she had been trained to perform her sequence, and I with a little coaching from Allen Gardner had learned quickly to perform my part of the act—a kind of chimp-and-professor act.

Anyhow, that game broke the ice. Moja offered me in her hand a piece of candy-coated gum Allen had given her. I took it and said *thanks* in signs. She kept on staring at me. (We were face to face and close, as I had hunkered down to match her height more closely.) She came closer, her eyes focused right on my chewing mouth. Allen quietly told me that Moja had not expected me to take the gum—just to take notice of and applaud her nice manners and say, *No, thank you.* Learning that, I pushed the gum out between my lips and let Moja take it from me with her delicate and agile lips. At that we were friends, and she invited me to have lunch with her; that is, she pulled me by the hand to her quarters and showed me where to sit, while she played games with her surrogate parent who was trying to get her to cooperate in getting cleaned up for lunch.

I have interacted with many animals from cats to horses and with human infants both normal and impaired, and without romancing or anthropomorphizing I can say that Moja acted more like a bright child than like any other young animal I have interacted with. I am not saying that she acted like a fellow creature, and she did not act like an actor playing a prelearned role—that is, there was clearly no apprentissage. If there was any dressage—one creature being directed by another's control—it was by Moja pulling the reins and me performing the amusing antics, which she seemed to be enjoying in much the same way that a toddler enjoys making Grandpa play horse or polar bear or tiger.

When I met Washoe, she was no infant, and any carefully fostered similarity to a human infant had long gone. She was a young but mature female chimpanzee just at the end of her estrous cycle and instinctively or hormonally hostile to primate females and interested in a vague way in primate males. However, she was primarily out to enjoy a walk or romp in the spring woods as a break from confinement in winter quarters.

We, that is, Roger Fouts, Washoe, my wife Ruth, and I, had reached the edge of the woods before Washoe took any notice of my signing. She stopped her impetuous progress at a fence and was looking out across a wide field. When she turned back to the rest of us, I caught her gaze and asked in signs *what look-at?* Washoe quickly put her left hand up to her forehead (her left side was nearer to us), formed a gesture that is *cow* in ASL, and looked back across the field again. I looked too and saw that there were cows at the far edge, just visible.

Next, Washoe was all playful chimp again and climbed and swung from a sapling. A little later she was following a wagon road, and I was walking parallel a few yards to her right. I picked a purplish wildflower and held it toward her, asking in sign, *what that?* Washoe made a quick gesture that could easily be interpreted as the sign *flower*, and picked it carefully out of my fingers. Then she ate it. A few moments later, she had swung around so that she was striding toward me from in front with outstretched arms. So much gratitude surprised me (it was only a small flower) but I held out my arms too; we embraced; her kiss was huge and wet.

That evening Roger complimented me on my quick thinking. I didn't know what he meant. It seemed that Washoe had not been showing friendship in that striding, arms-out walk but making an aggressive threat display; the next thing could have been a blow, but I had misinterpreted the display and my hug had turned her aggressive display to affection.

Once again I saw no evidence of long, well-formed sentences; only a few signs. However, in both encounters I found animals that interacted in a playful and humanlike way and not at all with the concentrated seriousness of a programmed or guided trained animal; nor was what they did much like the behavior of a friendly but quadrupedal domestic animal accustomed to human interaction—but this could be an obvious effect of bipedal gait, an eyes-front face, and humanlike arms and hands. Of course I second the suggestion by Sebeok and Umiker-Sebeok that the full expertise of animal training methods is needed, but so is the empathy of ethologists like Lorenz and zoologists like Griffin. There is just too much on the outside of language for Premack, Gardner and Gardner, Rumbaugh and Savage-Rumbaugh, Terrace, and others to define it fully. Their methods cannot account for all the behavior that falls outside their definition of language. The fault is as much that of linguists as of psychologists; for linguistic behavior has to grow out of communicative behavior. The two-way communication between apes and human beings in these experiments is rich and interesting, and it is important material for a science of humanity. I think that there is much that we can learn about learning and teaching from apes, even now.

Is There Language?

Consider three categorical statements. All three may seem bizarre, but I offer them seriously. The three come in order, the easiest one to accept first:

(1) No one can teach an ape language. (All the negative critics agree with that.)
(2) No one can teach anyone language. (Human beings are born with the capacity or compulsion to acquire a first language.)
(3) No one can learn or needs to learn language. (See below.)

Linguists with a strong rationalist bent believe that a capacity for language, a language acquisition device, or a "language organ" is inborn, and therefore that human infants need only exposure to some language in use—that children do not need to learn "language" but only a specific set of language habits. I cannot agree completely with such a view, but I do believe no one can learn language— not because we are human and prewired for language, but because in all the world there is no such thing as language. "Language," to begin with, is a word. This word with no article or other modifying word in front of it may stand for a concept, a very abstract concept, perhaps the most abstract concept our minds can conceive. For example: "In the beginning was the Word, and the Word was with God, and the Word was God." I will not try to explain that, but I think all will agree it is abstract and profound. Notice also that the quotation refers to "the Word" not "word" but "*the* Word." That must mean something.

Of course when people use the naked word "language," they may mean something real and use "language" unadorned just to save time. Thus, in the statement, "The human species uses language; other species do not use language," the real meaning is "every human being belongs to a group that uses some specific language."

There is nothing wrong with that kind of shortening; using the word "language" like that is a shorter way of saying "all the languages ever used by humankind." However, using the word alone sometimes leads users into a trap, the wrong mental habit of thinking, "if there are languages, then there must be language in the abstract"; but there is not and there cannot be any such thing.

Another use of the naked word "language" seems to have more standing: the use of it to mean all the grammatical rules and semantic rules and sound-pattern rules that all languages have in common. The trouble with that usage is that for a long time it left signed languages outside the definition. Some thinkers would say that when all the specific things that do not belong to all languages are left out, what we end up with is not a description of universal language but of human cognition or thought. Leaving such arguments to philosophers and psychologists, we can conclude again that every member of our species can speak or sign at least one language but that no one can speak "language" or sign "language" or learn "language" or teach "language." Language in the abstract is too abstract to handle with any organ except the mind, and we have no proof that the mind is an organ. Mind is more likely to be, as Bateson (1979) thinks, the nature of animal nature.

When I teach my students this fundamental principle of linguistics and epistemology, I use the concept and term "ball games." I ask them to think of three or four ball games; any ones will do. A class is usually able to come up with two or three dozen in a few minutes. We list them on the board. Then I ask if anyone in the class has ever played a ball game. Most say "yes," but I challenge them. In every instance, without exception, they were playing basketball, or table tennis, or hockey, or something else. It is handy to be able to sum up all these in the locution "ball games," but as long as people know what they are doing they cannot really be playing "a ball game." The concept "ball game" and any specific game are different logical types. So, too, "language" (if it is not an empty word) belongs to a different logical type from that of any specific language. Human beings can learn a specific language or languages once they are grown, but they cannot learn "language." "Language" may be the product of a specific human organ as Chomsky says, but I have to be shown. I have never seen or heard or felt or smelled or tasted or intuited langauge, and so I conclude that it has no existence except in abstract thought. English or ASL or Latin or several other languages I either can use comfortably or else have a hard time with—those I believe in.

Logical Types

The chimpanzees with signing human companions or teachers may be in reach of a real language, but it is far easier for most human signers who are not deaf and not raised in a signing family to use manual signs to stand for the words they are speaking or would be speaking if not constrained to be silent. Thus, the language the chimpanzees may be in reach of may be ASL or, much more likely, it may be English expressed in manual signs. The latter case makes the experiments more difficult. There is no clear evidence that deaf children acquire English by being exposed to manual signs used where spoken words are normally. (The Swedish researchers Ahlgren and Bergman, 1980, have found, however, that children exposed to signed Swedish acquire neither Swedish nor Swedish Sign Language, from which the word surrogate signs are taken.) It is important, therefore, that the experimenters distinguish carefully between ASL and English partially expressed in manual signs.

It is clear throughout Terrace's book that Nim, with 60 different teachers, only one of whom was deaf, had something other than ASL as the language in his environment. Washoe was companioned while quite young with deaf native signers. Some of the chimpanzees Fouts worked with seemed to be able to understand spoken English words and to respond when requested by enacting the manual sign learned as gloss for the word (Fouts & Mellgren, 1976).

All such skills in chimpanzees and other apes must of course be tested against a thorough knowledge of the limits of dressage and appretissage. Before concluding that an animal has learned to do certain things with some of the structures of a particular language, things that are comparable with what a child of a specified age does (which I take it is all that Gardner and Gardner have concluded), it is necessary to be sure that the behavior cannot be explained nonlinguistically as apprentissage or dressage. If, as I think has been demonstrated, a chimpanzee can again and again respond appropriately in a creative, not a closed, way to a complex changing human environment, then that chimpanzee must be doing something significantly different from what is done by pigeons and apes in cages.

To summarize, I find that too many of the apes were two or three or more logical levels away from any real language. In addition, some of the experimenters consistently confuse "language" in the abstract with actual use of a specific language accessible to apes' receptive and productive capabilities. However, I find that the critics who attack the experiments have failed to provide any solid basis for denying what the animals have demonstrated. If an ape has behaved in a situation just as a child acquiring a language should behave and the information it receives and acts on is truly nonverbal, nonlinguistic, and nonmanual (that is, if the signs it sees are ignored and if like Clever Hans it gets other signals), then the whole matter of communication and language sciences is in a parlous state indeed.

I agree with Sebeok and Umiker-Sebeok and others who think it foolish to experiment with apes and "language," but only because, as aforesaid, I can see little profit in speculations about language in the abstract when so many of the specifics have escaped us. I do think that studies of systems of signification, within and across species, are needed. So too are studies of systems of communication, also across species and in a continuum from animal communication to communication in specific languages, signed as well as spoken. It would be a serious and ancient mistake to put up an artificial barrier between signification and communication in animals and signification and communication in that other animal. Like the Frenchman I can shout, "Vive la différence," but as one anxious to learn, I would add, "Thank God for the similarities."

References

Ahlgren, I., & Bergman, R. Paper presented at the First International Symposium on Sign Language Research, The Swedish National Association of the Deaf, Leksand, Sweden, 1980.

Bateson, G. *Mind and nature: A necessary unity.* New York: Doubleday, 1979, Chap. 4.

Eco, U. *A theory of semiotics.* Bloomington, Ind.: Indiana University Press, 1976, 8-9, 32-47.

Fouts, R. S., & Mellgren, R. Language, sign and cognition in the chimpanzee. *Sign Language Studies,* 1976, *13,* 319-346.

Gardner, R. A., & Gardner, B. T. Two comparative psychologists look at language acquisition. In K. E. Nelson (Ed.), *Children's language* (Vol. 2). New York: Halstead Press, 1980, pp. 331-369.

Lieberman, P. *On the origins of language.* New York: Macmillan, 1975.

Maestas-y Moores, J. Early linguistic environment. *Sign Language Studies,* 1980, *26,* 1-13.

Rosenthal R. (Ed.). *Clever Hans (The Horse of Mr. von Osten) by Oskar Pfungst.* New York: Holt, Rinehart & Winston, 1965.

Sebeok, T. A. *The sign and its masters.* Austin: Universtiy of Texas Press, 1979.

Sebeok T. A., & Umiker-Sebeok J. (Eds.). *Speaking of apes.* New York: Plenum Press 1980.

Stokoe, W. C. The classification and description of sign languages. In T. A. Sebeok, (Ed.), *Current trends in linguistics* (Vol. 12). The Hague, The Netherlands: Mouton, 1974, pp. 346-371.

Stokoe, W. C. *Sign language structure* (Rev. ed.). Silver Spring, Md.: Linstok Press, 1978.

Terrace, H. S. *Nim*. New York: Knopf, 1979.

Timaeus, E. Some non-verbal and paralinguistic cues and mediators of experimenter expectancy effects. In M. von Cranach & I. Vine (Eds.), *Social communication and movement*. New York: Academic Press, 1973, pp. 445-464.

CHAPTER 11
Prospects for a Cognitive Ethology

Donald R. Griffin

What, if anything, do animals think about? Do they have anything at all comparable to mental experiences as we know them? Are any of them ever aware of themselves, their surroundings, or the results likely to follow from their behavior? It would of course be manifestly absurd to overlook or underestimate the enormous differences in complexity, versatility, and range of comprehension between human and animal thinking (if the latter occurs at all). However, our direct knowledge of our own mental experiences, imperfect as it may be, is apparently quite sufficient to establish the importance of such experiences, at least in one species. Mental experiences will not wither away merely to keep science simple and tidy. Hence it is appropriate to inquire whether or not the differences between men and animals in this respect are qualitative and absolute, and to explore the possibility that particular animals may experience thoughts and feelings of one sort or another under various conditions. Determining the nature and extent of animal thinking is crucially important in our attempt to understand human uniqueness and our place in the universe, as has been discussed by Popper and Eccles (1977) and many others.

Animals often behave in ways that suggest anticipation of what is likely to happen in at least the near future. Much of their behavior seems to involve intentional choices among alternative actions of which they are capable. Most people at all familiar with complex animals tend to assume that in some situations such animals know what they are doing. Nevertheless, behavioral scientists have come to ignore the possibility that mental experiences, awareness, intentions, or the like might occur in any nonhuman animal. This reluctance is related to the widespread belief that it is impossible to identify or analyze mental experiences by any sort of independent observation. Even in other people they are held to be "private data," reportable only imperfectly and unreliably even by means of language. Furthermore, language is almost universally assumed to be a unique human attribute. Since strict behaviorists reject even the mental experiences of man as being inappropriate for scientific consideration, those of animals have appeared to be doubly out of bounds. (For the history of these ideas, see Schultz, 1975.)

If there were no continuity and commonality between human and animal thinking, the widespread use of animal surrogates to investigate many attributes of human thinking would appear to be seriously compromised—unless one were persuaded that even human mental experiences are of no consequence. For example, while no one would claim that the learning abilities of rats or pigeons approach the human level, it is tacitly assumed that general principles apply sufficiently well to both human and animal learning to justify significant comparisons. Indeed, educational policies have been formulated partly on the basis of experiments with animals. It may be argued that learning can occur in the absence of any conscious awareness of the relationships that are learned (Roland, 1978). However, awareness of learning is certainly one important component of conceptual thinking. Comparative psychologists have also studied problem solving and concept formation in animals with the important result, among others, of demonstrating that animals can deal successfully with simple concepts such as oddity or middleness and that they can solve problems that involve understanding simple relationships, or at least reacting appropriately to them (see, for example, Herrnstein, Loveland, & Cable, 1976; Medin, Roberts, & Davis, 1976; Premack, 1976; Riopelle, 1967; Thorpe, 1963). In addition, experiments by Gallup (1977), to be discussed later, provide strong evidence of self-awareness in chimpanzees.

Thinking is clearly not a unitary phenomenon or process, and therefore no one should expect it to occur in neat all-or-nothing packets. An inquiry into the possibility of animal thinking must thus be concerned with questions about which, if any, of the wide range of activities we loosely categorize as "thinking" occur in other species. In this difficult and challenging area it is unwise to rely on dogmatic assertions, no matter how familiar and customary they may be. What we need is new information to reduce uncertainty and improve understanding. Can this be obtained, and if so, how? One necessary, but perhaps not sufficient condition for the occurrence of awareness and thinking is the presence in the brain of patterned images or representations of outside objects and events, together with their temporal and spatial relationships. Whether such images are strictly iconic or are coded in any of a variety of noniconic ways is not important in this connection; probably both categories occur. Mental images are similar in many ways to perceptions, but they sometimes occur in the absence of current sensory input. (If one is uncomfortable with the term "mental," the term "internal representations" is almost equally appropriate, provided one recognizes that these are often dynamic and adaptable according to the animal's situation and needs.) Although mental images are considered meaningless by many behaviorists, they have been analyzed in considerable detail by psychologists (Nicholas, 1977; Segal, 1971; Sheehan, 1972; Shepard, 1978). A scientific journal called *Mental Imagery* has recently been established. While the existence of a technical journal does not rigorously demonstrate the reality of its subject matter (witness parapsychology, for example), it does testify to a substantial interest.

While thinking and awareness obviously involve many heterogeneous entities and processes in addition to mental images, the latter constitute particularly important components in mental experience. They also offer a promising starting point for a scientific approach to cognitive ethology, because hypotheses about their existence and their properties may be testable, as discussed below. Intentions are also of interest because it may be possible to gather verifiable evidence concerning their occurrence. I have recently suggested that it is time for a new look at these old problems (Griffin, 1976, 1977a, 1977b), and the approach that seems most promising is concisely summarized in the following terms:

> The communication behavior of certain animals is complex, versatile, and, to a limited degree, symbolic. The best-analyzed examples are the dances of honeybees and the use of American Sign Language by several captive chimpanzees. These and other animal communication systems share many of the basic properties of human language, although in very much simpler form.
>
> Language has generally been regarded as a unique attribute of human beings, different in kind from animal communication. But on close examination of this view, as it has been expressed by linguists, psychologists, and philosophers, it becomes evident that one of the major criteria on which this distinction has been based is the assumption that animals lack any conscious intent to communicate, whereas men know what they are doing. The available evidence concerning communication behavior in animals suggests that there may be no qualitative dichotomy, but rather a quantitative difference in complexity of signals and range of intentions that separates animal communication from human language.
>
> Human thinking has generally been held to be closely linked to language, and some philosophers have argued that the two are inseparable or even identical. To the extent that this assertion is accepted, and insofar as animal communication shares basic properties of human language, the employment of versatile communication systems by animals becomes evidence that they have mental experiences and communicate with conscious intent. The contrary view is supported only by negative evidence, which justifies, at the most, an agnostic position.
>
> According to the strict behaviorists, it is more parsimonious to explain animal behavior without postulating that animals have any mental experiences. But mental experiences are also held by behaviorists to be identical with neurophysiological processes. Neurophysiologists have so far discovered no fundamental differences between the structure or function of neurons and synapses in men and other animals. Hence, unless one denies the reality of human mental experiences, it is actually parsimonious to assume that mental experiences are as similar from

species to species as are the neurophysiological processes with which they are held to be identical. This, in turn, implies qualitative evolutionary continuity (though not identity) of mental experiences among multicellular animals.

The possibility that animals have mental experiences is often dismissed as anthropomorphic because it is held to imply that other species have the same mental experiences a man might have under comparable circumstances. But this widespread view itself contains the questionable assumption that human mental experiences are the only kind that can conceivably exist. This belief that mental experiences are a unique attribute of a single species is not only unparsimonious; it is conceited. It seems more likely than not that mental experiences, like many other characters, are widespread, at least among multicellular animals, but differ greatly in nature and complexity.

Awareness probably confers a significant adaptive advantage by enabling animals to react appropriately to physical, biological, and social events and signals from the surrounding world with which their behavior interacts.

Opening our eyes to the theoretical possibility that animals have significant mental experiences is only a first step toward the more difficult procedure of investigating their actual nature and importance to the animals concerned. Great caution is necessary until adequate methods have been developed to gather independently verifiable data about the properties and significance of any mental experiences animals may prove to have.

It has long been argued that human mental experiences can only be detected and analyzed through the use of language and introspective reports, and that this avenue is totally lacking in other species. Recent discoveries about the versatility of some animal communication systems suggest that this radical dichotomy may also be unsound. It seems possible, at least in principle, to detect and examine any mental experiences or conscious intentions that animals may have through the experimental use of the animal's capabilities for communication. Such communication channels might be learned, as in recent studies of captive chimpanzees, or it might be possible, through the use of models or by other methods, to take advantage of communication behavior which animals already use.

The future extension and refinement of two-way communication between ethologists and the animals they study offer the prospect of developing in due course a truly experimental science of cognitive ethology. (Griffin, 1976, pp. 103-105)

Because these questions are of such fundamental importance, it is worthwhile to clarify some issues about which unnecessary confusion has arisen, to extend the constructive discussion of cognitive ethology that has been rekindled, and to inquire how cognitive ethologists might investigate these old questions with new procedures. The basic goal of this sort of cognitive ethology will be to learn as much as possible about the likelihood that nonhuman animals have mental experiences, and insofar as these do occur, what they entail and how they affect the animals' behavior, welfare, and biological fitness.

Defining Awareness and Thinking

One reason for avoiding a cognitive approach to animal behavior has been the difficulty of agreeing on definitions of the terms and concepts involved. It is important to recognize at the outset that almost any concept can be "quibbled to death" by excessive insistence on exact operational definitions (Kupfermann & Weiss, 1978). Such widely used and clearly useful terms as "hunger," "memory," "aggression," or even "metabolism" have been subjected to erudite analyses in a search for definitions that will satisfy all demands and avoid every possible ambiguity. These efforts tend to come in waves, each followed by a truce of sheer exhaustion—after which the term continues to be used, but with clearer appreciation of the breadth of its connotations. Excessive concern to avoid all terms that cannot be rigorously defined suffers from the danger of retaining only verbal corpses displaying rigor mortis.

How then can we arrive at even approximate working definitions adequate to allow us to study the question of animal awareness without excessive confusion and misunderstanding? We can begin with the accepted usage of scholars working in other fields. The Random House *Dictionary of the English Language* defines *awareness* as "having knowledge, being conscious or cognizant, informed, alert"; *consciousness* as "awareness of one's own existence, sensations, thoughts, surroundings, etc."; and *mind* as "the element, part, substance, or process that reasons, thinks, feels, wills, perceives, judges, etc." A cognitive ethologist may inquire to what extent these attributes are present in various animals. We can also turn to the philosophers, who have been deeply concerned with the nature of minds. Edwards (Edwards & Pap, 1973), when he introduced the mind-body problem in a textbook of philosophy, defined *mind* as follows: "Feelings, sensations, dreams, and thoughts are the sort of phenomena which are usually classified as 'mental.' In calling them mental, philosophers usually mean that, unlike physical objects, they are 'private' or directly knowable by one person only."

Schaffer (1975) defined *mind* as follows: "as the term is used more technically . . . and in the philosophy of mind today, [it] encompasses sense perception, feeling and emotion, traits of character and personality, and the volitional aspects of human life, as well as the more narrowly intellectual phenomena." Elsewhere Schaffer stated: "One thing that sharply distinguishes man from the rest of nature is his highly developed capacity for thought, feeling, and deliberate action. Here and there in other animals, rudiments, approximations, and limited elements of this capacity may occasionally be found; but the full-blown development that is called a mind is unmatched elsewhere in nature." A cognitive ethologist may wonder whether perhaps the mental capabilities of animals will turn out to be more substantial and significant than we have been accustomed to recognize. Defining mental experiences as uniquely human discourages inquiry into the possibility of their occurrence in other species.

Kenny, Longuet-Higgins, Lucas, and Waddington (1972) devoted a lucid, thoughtful, and stimulating series of Gifford Lectures at Edinburgh to *The Nature of Mind*, without explicitly defining the terms "mind" or "mental." Ryle (1949), in a very influential book entitled *The Concept of Mind*, also avoided any specific definition of the term. In a second series of Gifford Lectures (Kenny, Longuet-Higgins, Lucas, & Waddington, 1973, p. 47), Kenny stated that "to have a mind is to have a capacity to acquire the ability to operate with symbols in such a way that it is one's own activity that makes them symbols and confers meaning on them." The communicative dances of honeybees certainly satisfy this criterion; for it is each forager's own activity that makes the waggle dance into a symbolic statement about distance, direction, and desirability of something the dancer has visited. (For detailed descriptions of the dance communication system see Frisch, 1967; Hölldobler, 1977; Lindauer, 1971; and the discussion of cognitive interpretations below.) Some philosophers may object to calling the bee dances symbolic on the ground that only thinking creatures can recognize symbols, so that use of the term *symbol* implies that bees do think, and thus tricks the reader into accepting the conclusion at issue. For the moment I mean to point out simply that the bee dances satisfy this particular definition.

Elsewhere in these Gifford Lectures, Longuet-Higgins offered quite a different sort of definition: "An organism which can have intentions I think is one which could be said to possess a mind [provided it has] . . . the ability to form a plan, and make a decision—to adopt the plan" (Kenny et al., 1972, p. 136). Many animals behave as though they do have plans of at least a simple sort and adjust their behavior appropriately in attempts to carry them out.

The neurophysiologist John defined consciousness as

> a process in which information about multiple individual modalities of sensation and perception is combined into a unified multidimensional representation of the state of the system and its environment, and integrated with information about memories and the needs of the organism, generating emotional reactions and programs of behavior to adjust the organism to its environment . . . . Consciousness about an experience is defined as information about the information in the system, that is, consciousness itself is a representational system. . . . Perhaps our philosophical quandary [concerning mind-brain dualism] arises from the assumption that organized processes in human brains are *qualitatively* different from organized processes in other nervous systems or even in simpler forms of matter. Perhaps the difference is only quantitative; perhaps we are actually not as unique as we have assumed. (Thatcher & John, 1977, pp. 294-304)

Conscious awareness and mental experience may sometimes be limited to a single sensory modality, vision, for example, so that a rigid requirement that consciousness entail integration across modalities may not be justified. In other respects, however, this definition is close to the cautious and tentative views of many neuroscientists, and it is important to note that it does not limit conscious awareness to our species.

Even though philosophers have not settled on a single explicit and generally acceptable definition of mind, it would be foolishly inhibiting to give up all attempts to understand whatever processes give rise to such concepts as awareness or intention until they can be defined with the assurance with which a chemist assigns a structural formula to a purified compound. When the nature of a sample is unknown, it would inhibit the chemist needlessly to insist that he consider only one of a set of well-defined molecules. The sample might contain something never before analyzed, or, more likely, some mixture of known and unknown substances. A comparable situation may confront us with regard to mental experiences. What now seems to be a single, although vaguely defined entity, such as awareness, may well turn out, when fully understood, to be a mixture of known or unknown processes. We will make little progress, however, if we throw up our hands in dismay and refuse to study the unknown sample at all.

Depending on how awareness is defined, the question of its possible existence in other species can vary all the way from the trivially obvious to the most preposterous level of implausibility. At the first extreme one might define awareness as any capacity for reaction; but this would allow the inclusion of all living organisms, and of even such simple mechanisms as a mousetrap. The distinction between awareness and responsiveness should not be overlooked. At another extreme one might demand the use of written language, or the most complex levels of understanding known to human thinkers—the creative insights of Beethoven, Einstein, or Whitehead, for instance; but these requirements would eliminate most of our own species.

A confident belief in biological evolution implies that, while mental experiences of other species may differ greatly from ours, they will turn out, when fully understood, to share important properties with the entities we meet through our individual introspection. Of course, this is a question to be kept open and under investigation, not one about which firm statements or dogmatic assertions are remotely appropriate. To put the question in slightly different terms, how do those results of brain activity that we call mental experiences differ among various species of animals? The attempt to study animal thinking was largely abandoned early in the 20th century, but so much has since been learned about animal behavior, and how it can be objectively analyzed, that a cognitive approach to ethology has much better prospects of success than in the days of Darwin and Romanes. In the early stages of such an effort it seems advisable to devote primary attention to a search for independently verifiable evidence that animals have mental images of objects, relationships, or events, and intentions about their own future activities, because this seems less difficult than devising tests of hypotheses about the subjective nature of "raw feels." These questions will be discussed in more detail in a later section.

Psychoneural Relationships

Most behavioral scientists, psychologists, and ethologists are thoroughgoing materialists. They believe, or at least operate on the working hypothesis, that the heterogeneous set of processes leading to what we call "mental states" depends directly upon complex activities of central nervous systems, especially interactions among various excitation patterns. Some small fraction of these patterns of interacting activity in human brains generate what we call "mental experiences." In the present state of its development, neurophysiology cannot determine whether there are significant qualitative differences between processes that are and are not accompanied by subjective mental experience. In a limited sense, the experiences may be related to brains as programs are related to computer hardware (a suggestion discussed by Longuet-Higgins in Kenny et al., 1972, p. 25). However, insofar as this analogy is valid, only a small fraction of the programs are accessible to conscious experience.

Postulating the existence and significance of mental states in other species implies little or nothing about their functional relationship to the central nervous system of the person or animal concerned. It makes little if any difference, for the purpose of this discussion, whether mental experiences are assumed to be identical to some pattern of neural activity, or whether one prefers one of the many other types of psychoneural relationship discussed in a recent symposium edited by Globus, Maxwell, and Savodnick (1976). (See also Puccetti & Dykes, 1978.) The question I am raising is not dependent on whether or not one is a materialist; it concerns the degree of similarity of mental experiences resulting from brain function in our species and others. Only if one postulates a major, qualitative difference between psychoneural relationships across species does this complex area of philosophical concern become directly relevant.

Of course, any evidence bearing on the mind-brain relationship is important in its own right, and insofar as it involves observable mechanisms it becomes pertinent to inquire whether these may in fact be different in human and other brains. Lateralization of control of linguistic communication is one of the clearest cases in point. This was once considered a unique feature of human brains and was often advanced as neurobiological evidence of a species-specific human mechanism closely related to speech, and hence by implication to conceptual thought (reviewed by Brown, 1976; Dimond & Blizard, 1977; Galaburda, Le May, Kemper, & Geschwind, 1978; Harnad, Steklis, & Lancaster, 1976; Neville, 1976). The localization of a certain function in one part of the brain, however, does not in itself tell us much about just what is occurring there. Many animal brains are slightly asymmetrical, and some are more so than the human brain. Especially important are the recent discoveries of Nottebohm (1977), which have demonstrated that in songbirds the control of vocalization is almost entirely concentrated in one-half of the brain (see also Corballis & Morgan, 1978). These cerebral asymmetries are much more pronounced than those found in the human brain. Should we therefore accept songbirds along with signing apes into our Select Kingdom of talkers and thinkers? Many otherwise normal people do *not* have speech-control centers that are demonstrably larger in one hemisphere (Galaburda et al., 1978); should they be banished from humanity? If we define human uniqueness on too narrow a foundation, we are in danger of having it undermined whenever the same feature is discovered in some other species.

Neurophysiologists studying computer-averaged evoked potentials from the brains of both human subjects and laboratory mammals have identified certain waveforms of relatively long latency (on the order of .3 second) that are much larger when the stimulus that elicits them is a signal recognized as important by the person or animal (reviewed by Thatcher & John, 1977). These potentials seem to be neurophysiological correlates of paying attention to a stimulus; but it would be premature to interpret them as physiological counterparts of conscious mental experience, although further developments along these lines may approach that long-sought goal. Meanwhile, it is significant that these electrical potentials appear so far to be basically similar in men and other mammals.

Ryle (1949) ridiculed the belief he ascribed to many philosophers that a person's mind is a "ghost in the bodily machine." Nevertheless, like Skinner (1957, 1974), he freely recognized that we do think and experience internal representations of external objects and events. The "ghost" that both Ryle and Lashley (1923) were trying to exorcise is derived from a belief in an extramaterial realm, a sort of mental universe inherently different in kind from the material, physical world. One can agree completely with Lashley and Ryle about this point and still ask meaningful questions about the thinking or experiencing of internal images in other species. The scientific consideration of animal awareness does not require any of the following: (1) that we ascribe anything approaching the human level of intellectual capacity to other species, (2) that we postulate immaterial mental essences, or (3) that only if we endow animals with immortal souls can we reasonably postulate that they are aware of themselves and their surroundings, as suggested by Humphrey (1977). One certainly need not depart so far from common sense and everyday experience as physicists have done in postulating antimatter. It seems most reasonable and parsimonious to postulate, tentatively and pending new evidence, that thinking and experiencing are related in comparable ways to the functioning of central nervous systems in various species. It contributes very little to our understanding of these difficult problems to erect and then demolish straw ghosts.

The Behavioristic Taboo

The more thoughtful behaviorists recognize the reality of mental experiences or mental states in human beings, but strongly urge that introspective reports about "mentalisms" be discounted and ignored as unreliable and meaningless. Although most behaviorists are careful to avoid dogmatic negative statements, to a significant extent behaviorism defines itself in negative terms, by seeking to exclude mental experience from scientific psychology. It is often fruitful, however, for critical scholars to approach a difficult topic from differing viewpoints; and therefore scientists who interest themselves in the mental experiences of other species need not be deterred by proclamations of aversion from those who do not. Extreme forms of behaviorism tend to become little more than irrelevant pleas of willful ignorance.

Methodological behaviorism is a more cautious and reasonable viewpoint, which emphasizes the great difficulty of gathering consistent and verifiable data concerning mental states, even in other people, let alone in animals. It has led psychologists into strict or radical behaviorism, as explicitly set forth in an important but neglected paper by Lashley (1923). Ethologists have accepted methodological behaviorism so thoroughly that their writings almost never consider the possibility that any nonhuman animal might know what it is doing, and this silence is so complete as to imply that animals never think at all. However, *rigorous* proof that something does not exist is notoriously difficult to provide. Methodological behaviorism justifies agnosticism rather than dogmatic denial of the existence or significance of what has seemed to lie beyond the effective reach of any scientific procedures yet developed.

Some behavioral scientists vigorously proclaim that they are not interested in animal awareness even if it does occur. Their antipathy sometimes seems to be so strong as to suggest that they really do not *want* to know about any thinking in which animals might engage. Nineteenth century astronomers would have been ill advised to devote a major effort to speculating about topographical features on the far side of the moon, since no available methods could test any hypotheses they might have developed; but this did not mean that there were no craters and mountains beyond their view, or that astronomers should have been inhibited from wondering about them and planning how they might be studied in the future. The study of mental processes in other species is in a somewhat comparable state, not because of spatial remoteness but because of more subtle difficulties of gathering data and of interpreting data already available or readily obtainable. The taboos of behaviorism also constitute a self-made obstacle we *can* eliminate.

Mackenzie (1977) reviewed the limitations of behaviorism, and the following quotations from Whiteley illustrate the widely held views of many thoughtful scholars:

> It seemed obvious to all philosophers before the present century that we do have private experiences and that we can inform one another about their character and be understood adequately if not with perfect precision. . . . When these obvious truths are denied, one feels oneself in the presence of that legendary monster the Metaphysician, raptly following the track of his *a priori* argument to the very doors of the madhouse . . . [this] thesis is as absurd as it looks . . . . contrary to what behaviourism teaches, our knowledge of people's experiences is often much more secure and dependable than any generalizations we can make about their behaviour.
>
> To understand why people act as they do, we need to know what goes on in their awareness; and for this we must sometimes be able to grasp their points of view, to share their insights, to imagine how they think and feel; and since much of our information on these matters comes from what they say, we must be able to understand their language . . . . many psychologists and philosophers champion a study of people which confines itself to publicly observable behaviour, ruling out their testimony as to unobservable states of consciousness—a study of man, which, following the human drama as it were in dumb show, deals with groans and smiles, fighting and copulation, but has nothing to say about pain and joy or love and hate. But it is unwise to let one's view of the nature of things be determined by the conveniences of some technique of investigation. (Whiteley, 1973, pp. 8, 22, 116)

Communicative Behavior as a "Window"

Human language is widely viewed as not only the best window through which we share our thoughts, but also as essential to thinking itself. Allied to this nearly universal belief is the corollary that only our species has the capacity for language, and hence for thinking of any but the most rudimentary kind. Much human thinking is conveyed nonverbally, however, by gestures, expressions, intonations, and other forms of communication that are largely filtered out on the printed page—largely, but not entirely, because many nonverbal nuances are often conveyed in writing, along with strictly matter-of-fact sentences; but to restrict our view of human communication to printed grammatical sentences is to ignore much of its most important substance. We often think hopefully, fearfully, threateningly, or affectionately about things, events, or people. Whereas most such thoughts can be translated into grammatical sentences, many of their internal representations in our brains or minds may have more in common with nonverbal than with verbal communication. Human thinking may involve the internal manipulation not only of something akin to words but also of processes that are closely related to nonverbal messages.

At this point ethology has made a contribution of fundamental importance by discovering a rich variety of nonverbal communication in many types of animals. This evidence has recently been reviewed by Smith (1977) and by the contributors to a monograph edited by Sebeok (1977). Messages of threat, submission, and courtship are clearly recognizable in many groups of moderately complex animals, including all classes of vertebrates from fish to mammals, insects, arachnids, crustacea, and cephalopod mollusks. When animals employ communicative behavior to exchange such messages, the possibility arises that they are thinking something roughly comparable to the message.

I have recently suggested that animal communication offers a potentially powerful method for attacking some of the challenging questions that seemed hopelessly inaccessible to the original behaviorists (Griffin, 1976). The suggestion is basically a very simple one: to the extent that language is the best available window through which we learn about human thinking, why not open the window somewhat wider by recognizing that nonverbal communication can serve the same basic function as words and sentences, and that it could be used to gather information about the thoughts of other species, as well as those of our fellow men?

Human verbal and nonverbal communication often seem to be entirely different processes: the moods and emotions that make up most of the latter cannot be translated with complete fidelity into words and sentences, and it would of course be enormously difficult to convey highly abstract ideas by gesture. This line of thought has led many scholars to conclude that since only we use verbal language, our mental experiences must be totally different from those exchanged via the various types of nonverbal signals employed by both animals and men. However, insistence on an unbridgeable dichotomy between these two categories of communication fails to do justice to the creative versatility of human language. Even though absolute accuracy of translation between verbal and nonverbal communication is even more difficult than between two languages, sentiments and emotions can be conveyed even in cold print. The mental experiences underlying and conveyed by verbal and nonverbal communication may well have enough in common that with sufficient time and effort many of their most important components can be transmitted by either type of communicative exchange. It is clear, nevertheless, that some mental states are more conveniently conveyed by gestures than by sentences, and vice versa. Furthermore, human language allows our thinking to be far more complex and versatile than that of other species. It is not universally agreed that all human thought depends totally on language, and the potential usefulness of animal communication as evidence about animal thinking does not depend on an assumption that thinking and communication are so tightly linked that one cannot occur without the other.

Ethologists have made a good start toward exploiting the possibility that communication could serve as one sort of window on animal thinking by describing and analyzing how various species do in fact communicate. A fruitful next step might be for investigators to use and extend this knowledge to devise experiments in which animal communication could be used to ask questions about any thinking that might be going on in the brains of the animals under investigation. This general goal might be approached through the use of models or motion pictures of gestural communication and playbacks of sounds or electrical signals from tape recordings. A beginning has been made with gulls by Galusha and Stout (1977) and with iguanid lizards by Jenssen (1970).

One form of objection to this suggestion questions that one can judge from communicative behavior whether or not the animal actually intends to communicate. Many interacting systems are known in which information is transferred from one unit to another with predictable effects on the activities of the receiving unit. The sensor of a heat-seeking antiaircraft missile transmits to the steering mechanism information about the location of hot exhaust gases, but this does not demonstrate an intention to destroy the target aircraft. The DNA of an *Escherichia coli* bacterium contains coded information that predictably affects the replication of daughter cells and their enzymes. A bacteriophage can inject other DNA which alters this information and causes different enzymes to be synthesized and replicated. Information is thus transmitted with predictable effects, but this is not sufficient to indicate that the bacteriophage understands what it is doing. Can a reasonable and meaningful distinction be drawn between bacteriophages and animals as complex as honeybees, songbirds, or chimpanzees? One important difference between self-guiding missiles and DNA, on the one hand, and complex animals, on the other, is that the former do not announce their intentions through communicative behavior, whereas the latter sometimes do. To the best of our knowledge, bacteriophages do not exchange messages with other organisms; they do not threaten, protect, or reassure their social companions.

Communication about mental images and intentions offers cognitive ethologists an important opportunity for experimental verification and analysis. The
communication of a message may involve any of three types of relationships,
alone or in combination: (1) the message may relate directly to the animal's
perception of the immediate situation, that is, it may report about current
sensory input; (2) the animal may report about information acquired at an
earlier time and stored as some sort of memory; or (3) the animal may announce
an intention or a plan for future behavior. The third category is especially
important, insofar as it occurs, because it may involve patterns of information
quite different from anything arriving contemporaneously through the animal's
sense organs or retrieved from its store of memories. To be sure, the elements
involved in a plan will in many cases have been sensed or perceived before, but
they may be recombined in novel ways when projected into the future. Thus, if
such intentions or plans are consistently reported, and verified by correspon-
dence with subsequent behavior, they provide a more convincing type of evidence for the existence of mental images than do responses to current, or even
to remembered past sensations.

One reaction to the suggestion that a cognitive ethology might be developed on the basis of communication by animals about their mental experiences is to assert that comparative psychology and ethology have for many
years been actively engaged in analyzing the minds of animals through investigations of their behavior, and in particular their abilities to learn complex
discriminations (Campbell & Blake, 1977; Mason, 1976). When an animal
learns to respond to some stimulus, the experimenter is in a sense communicating with the animal and learning something about what goes on in its brain or
mind. Almost everything discovered by psychologists and ethologists about
animal behavior could be described in terms of postulated thoughts, intentions,
and mental images. But to do so tends to force a choice between two important
assumptions: (1) that all responses, or at least all moderately complex learned
responses, entail awareness on the animal's part of the relationship on the basis
of which it responds appropriately, and (2) that there is no significant difference
between executing a complex learned response and being aware of the relationship that has been learned.

The highly significant distinction between responsiveness and awareness
is of course familiar to all of us, but the difficulty of making the distinction
from behavioral evidence alone can be illustrated by a hypothetical example. Suppose an enemy air force includes two types of rocket-propelled antiaircraft missiles that are identical in size, speed, and all other features we can
observe from a distance; both overtake and destroy our aircraft (and themselves
as well) by means of identical pursuit maneuvers. One type is steered by a
heat-seeking guidance system and the other by a kamikaze pilot. If a radio link
were available, we might hear the pilot say something that would convince us of
his existence and awareness of his mission. Of course he might hold his tongue
right to his fiery end; so that absence of communication, or our failure to

intercept it, would not prove his nonexistence. Communication by voice or other means, however, might well convince us that a live, conscious pilot was flying one machine. Of course, the air force might wish to deceive its enemies by installing a radio and tape recorder that would transmit patriotic exhortations; but if a two-way communication link were available, we might well be able to distinguish such a contrivance from a real pilot by carrying on even a simple dialogue.

The theoretical advantages of communication as a window on animal minds do not mean that this is the *only* sort of evidence that should be considered. Obviously, much animal behavior is strongly indicative of awareness and at least simple kinds of thinking. However, a major and constructive contribution of 20th century psychology is the recognition that even complex behavior need not be accompanied by conscious experience. Behavior consistently predicted by prior communication of intentions, however, can provide strong evidence that people or animals have some awareness of what they are doing or are about to do. When animals do something equivalent to asking questions, especially questions about future activities, this provides strong evidence of some sort of internal representation or dynamic imagery of future events. Chimpanzees using gestures derived from sign language or other forms of symbolic communication respond appropriately to questions such as *What is the name of* . . . (Bourne, 1977). They also seem at times to reverse the process and themselves ask questions by the same communication methods. To be sure, Sebeok (1977, pp. 1067-1069) emphasized that Clever Hans errors, or the emission of unrecognized signals by human experimenters, may contaminate many of the experiments in which apes seem to be communicating symbolically. However, the number and variety of such experiments has now reached the point where it seems over-cautious to assume that they could all be based on such errors. Future extensions of these experiments may well yield stronger evidence for or against the occurrence of mental images and intentions in apes.

Many social insects, including not only honeybees but many of the wasps and ants, engage in solicitation behavior that results in regurgitation of material from the stomach of a returning forager. This provides the opportunity to receive chemical information about what a forager has been gathering (reviewed by Wilson, 1971). Should we interpret such behavior as asking a question in the form of *What have you brought?* The custom of behavioral scientists has been to avoid any such notion; but, as in so many other instances of this sort, we really have no convincing evidence pro or con. When a honeybee dancing on a swarm cluster stops dancing about the cavity she has visited and begins to pay attention to the dances of one of her sisters, could this change in behavior reflect the intentional asking of a question, *How good is the cavity you have visited?* While such interpretations cannot yet be tested rigorously, they deserve careful consideration, and it would be desirable to gather evidence about their plausibility.

The carefully argued views of Taylor (1962) are helpful in placing into a reasonable perspective the possibility of learning something about animals' awareness and thinking by means of information conveyed in their communicative behavior. In a real but limited sense this would require the use of introspections, and many will recoil in horror from the suggestion that a source of data long ago rejected by many human psychologists might nevertheless be useful with animals. However, as Taylor explained, in order to analyze the nature of human perceptions

> it is necessary to get subjects to report their acts of knowing as they occur, thereby making their private events public. This does not involve introspection in the sense in which Wundt and Titchener understood it. The subjects are asked to report not on what is going on in their minds simultaneously with the act of knowing, but simply on what they perceive at the moment . . . there is no need to postulate some mysterious form of neural energy . . . if consciousness can be explained without requiring any neurons to violate the law of their own nature, we may at last register a claim for the recognition of the study of conscious experience as a legitimate branch of natural science. Thus the revolt of the early behaviorists against the "mentalist" assumptions of classical psychology has led, through the assiduous investigation of the laws of behavior, to a new assertion of the scientific respectability of the once despised phenomena of consciousness. (Taylor, 1962, pp. 353-363)

Social Institutions

An important idea developed by Searle (1969) has considerable relevance for these questions. Searle distinguished "brute facts" from "institutional facts." The former are physical or mental data or relationships that do not depend on social rules, for example: *The bird landed on a horizontal branch*; $E = IR$; or *My tooth hurts*. Institutional facts, on the other hand, are exemplified by such statements as *The home team won by a double with bases loaded in the bottom of the ninth inning*; *The tax court disallowed his claim that the weekend in Las Vegas was a necessary business expense*; or *A honeybee with her stomach full of two molar sucrose recruited 77 of her sisters by waggle dances oriented 90 degrees left of vertical*. In examples such as the last three, a description confined to physical or mental events does not suffice to explain what has happened; one must also know the social rules or institutions that provide a framework within which the specified activities become meaningful. The inclusion of the zoological example in this list of institutional facts is justified because the dance communication system of honeybees constitutes an institution in Searle's sense.

There is no doubt that many social animals communicate by systematic codes that convey information and often lead to predictable changes in the behavior of the animal receiving the message (Sebeok, 1977; Smith, 1977). To a human observer the behavior is much more understandable when he has learned the code; by reading the communication signals he can predict how the receiver will behave. Such predictability is not perfect, but it is far better than random guesswork, and the more fully an ethologist understands the whole social situation, the more accurate the prediction. Nevertheless, hard-nosed critics are sure to object that there is a crucial difference between human social institutions, which are understood by the participants, and animal communication, whose rules, it is implied, resemble the activities of self-guided missiles or bacteriophages.

The fact that animals behave in a rule-guided fashion in their social interactions with conspecifics suggests, but does not prove, that they are aware of the rules or, in Searle's terms, the institutions that give meaning to their behavior. Often the behavior in question, such as emitting a particular pattern of sound, raising one set of feathers rather than another, or assuming a certain posture out of a large number within the animal's capability, would otherwise be trivial. We can employ the criterion of reportability. An articulate baseball player can describe the rules of the game to someone who does not know them; yet many institutions, including games, are understood in an intuitive fashion, and it requires some intellectual effort to articulate the rules. Few people have ever read the laws of their states, but we usually know whether we are breaking them.

Can any other species report the rules governing its social institutions? Not, as far as we know, in explicit, verbal fashion. Are we therefore obliged to assume that no animal mother intentionally helps her youngster learn the rules? Must all young animals either have the rules genetically encoded or learn them entirely by observation plus trial and error? We cannot answer such questions very well on the basis of available data, but pertinent evidence might well be obtainable by appropriate research in cognitive ethology.

An important distinction has been made by Ryle (1949) and others between knowing *how* and knowing *that*. As applied to animals, it is widely held that while they may know *how* to perform complicated patterns of behavior requiring discriminations between moderately complex stimuli, they do not have any understanding *that* they are performing such actions as aggressive displays or nest building. This distinction is a difficult one to analyze except through reports by the organism itself. It is on this basis that we make judgments about our fellow men and women as to whether they know *that* as opposed to knowing *how*. If a person wishes his companions to think him ignorant of what he is doing, he may make it very difficult to discern whether he merely knows *how* or also knows *that*. Does a dancing honeybee know *that* food or a suitable cavity is located at a certain distance and in a certain direction, or does she only know *how* to recruit her sisters to this location? What would be required for an animal to convince us that it knew *that* rather than merely *how*? Presumably an ape with an adequate vocabulary in ASL might describe what it was doing and thus provide convincing positive evidence. But are we justified in ruling out altogether the possibility that animals performing complex discriminations and choosing among several possible patterns of behavior the one most appropriate to the situation are necessarily unaware *that* this behavior will produce certain results? As in so much of this challenging area, open-minded agnosticism seems appropriate.

Testable Hypotheses Concerning Cognitive Ethology

Speculations about the possibility that a given sort of mental experience may occur in a particular animal need not, and should not convert a scientist into an advocate searching for evidence to support a prior conviction rather than an investigator testing a working hypothesis. It is of crucial importance to keep an open mind and be fully prepared to give appropriate weight to either negative or positive evidence. One objection to the ideas I am discussing is the claim that no testable, falsifiable hypotheses are generated by postulating animal awareness (Campbell & Blake, 1977; Krebs, 1977). This viewpoint, central to the behavioristic position, entails the assumption that no matter what an animal may do, it will be possible to interpret its behavior equally well without assuming any awareness or mental imagery (Haugeland, 1978). This assumption, however, can be readily tested in the case of other persons. Because we all share the belief that we *do* have thoughts, plans, intentions, memories, and anticipations, we can often predict the future behavior of a speaker on the basis of what he says. Such predictions are tested, and verified or falsified, so commonly that the whole matter seems too obvious to require explicit statement. Yet behaviorists reject introspective reports because on close examination these often prove less than perfectly consistent or reliable as predictors of future behavior. When we are dealing with biological phenomena as complex as behavior, however, few data are ever 100% consistent or reliable. We have learned to make good use of less-than-perfect information to improve our understanding of many other biological processes. Why is perfection so adamantly demanded in this case?

The standard response of a behaviorist is to insist that only external contingencies of reinforcement are of any interest, at least to him; but we are all free to form our own value judgments as to what is or is not of intellectual interest (although the behaviorist might argue that our choices are only illusions, and that external contingencies have determined what we will say we are choosing). This behavioristic disclaimer of concern with mental experiences is, of course, a minority view among people as a whole. And even in the academic fields of psychology, philosophy, and the social sciences many scholars have never succumbed to this form of intellectual myopia, or else are well on their way to recovery (Mackenzie, 1977). A surprising number of ethologists, however, seem to lag far behind the current swing of the pendulum away from positivism in general and radical behaviorism in particular.

One significant type of experiment in cognitive ethology was developed by Gallup (1975, 1977), and the results provide clear evidence of self-awareness in chimpanzees. These and other great apes have often learned to use mirrors to examine parts of their bodies that they cannot see directly. Gallup gave one group of chimpanzees ample opportunity to use mirrors, while others had no such experience. Then, under deep anesthesia, a conspicuous spot of quick-drying inert material was placed on the forehead or earlobe. Upon awakening, the chimpanzees paid no attention to the markings, indicating that no local tactile stimulation was present. When a mirror was provided, however, the chimpanzees who were familiar with their mirror images immediately reached for the colored spot and rubbed it or picked at it with their fingers, thus demonstrating that they recognized the mirror image as a representation of their own bodies. Those lacking experience with mirrors did not react in this way.

Gallup's type of experiment has so far yielded positive results only with great apes. Despite intensive efforts, gibbons, monkeys, and other laboratory animals have failed to react to mirror images as replicas of their own bodies. Gallup therefore has concluded that no other animal has the capacity for self-awareness. Should the capability of responding to mirrors appropriately in Gallup's experimental situation be equated with the possibility of self-awareness, or would other tests prove more appropriate for other species and provide evidence for the concept of oneself? Only further investigation can answer such questions; meanwhile, Gallup's experiments provide a clear and successful example of a well-controlled, objective, verifiable experiment in cognitive ethology.

In considering the possibility that animals may be capable of self-awareness, it is instructive to separate this question from the larger question of whether they have any awareness at all. If not, then the question of self-awareness obviously does not arise. Let us therefore explore the first possibility, that animals are capable of *some* types of awareness but not of self-awareness. This means that we assume the animal is aware of its companions and their actions and communication signals, as well as of its own physical surroundings, the ground beneath its feet, the wind that blows against its skin, and so forth, but that it is unable to entertain the concept of "selfness." However, an abundant flow of sensory input is always arriving at the animal's central nervous system from its own body—so that we must postulate that this input is somehow selectively barred from reaching the animal's awareness.

Although this type of "awareness of everything but me" is conceivable, it becomes increasingly less plausible the wider a range of awareness the animal is postulated to have of its inanimate and animate surroundings. If we allow a particular animal to be aware of a reasonably wide range of objects, events, and relationships in the world around it, while denying the possibility of self-awareness, we run the danger of redefining self-awareness in a roundabout way, as a sort of perceived hole in the universe. Self-awareness and awareness of future death have been wid ly held to be absent from all species but our own (Popper & Eccles, 1977; Premack, 1976); but direct evidence is almost nonexistent, and therefore these questions constitute an appropriate challenge to cognitive ethologists to seek relevant evidence.

Postulating awareness is no barrier to the formulation of falsifiable hypotheses about human mental images and human behavior. Behavioristic ethologists, however, may wish to dig in their heels, figuratively speaking, and claim that matters are different with species incapable of linguistic communication. This brings us full circle to the importance of animal communication for the questions under consideration. Insofar as such communicative behavior serves to report about mental experiences, awareness, intentions, and the like, it is potentially capable of serving the same basic function as human speech. Obviously, vocabularies are likely to be much smaller, and the capabilities of the communication channels will probably be severely limited in comparison to human language. However, the progress of Gardner and Gardner, and others who have followed in their lead, makes it prudent to reserve judgment as to the limits that may eventually be reached by the great apes in communicating their thoughts to human experimenters. Unless we apply a sort of double standard, accepting verbal reports from our conspecifics but rejecting even the most consistent and reliably informative communication about animal intentions, it should be possible to gather and critically evaluate evidence that will serve to increase or decrease the likelihood that cognitive interpretations are appropriate.

It is not unduly difficult to outline experiments that might yield significant evidence of conscious intention on the part of animals capable of moderately complex communication. Chimpanzees exchange signals effectively even without special training (Marler & Tenaza, 1977; Menzel & Halperin, 1975; Menzel & Johnson, 1976). Unfortunately, it has proved very difficult to decode these natural systems of communication used by apes, so that we know very little about their scope and versatility. The recent studies of Savage-Rumbaugh, Wilkerson, and Bakeman (1977), however, reveal that gestural communication can be used spontaneously by pigmy chimpanzees to convey information about motions a male wishes a female to make, or positions he wishes her to assume, in preparation for copulation. Great apes have also been trained to exchange quite complex information by means of symbolic gestures, artificial tokens, or keyboards in the well-known experiments of Fouts and Rigby (1977), Gardner and Gardner (1975), Premack (1976), and Savage-Rumbaugh, Rumbaugh, and Boysen (1978), and those of the several investigators whose work is described in books edited by Rumbaugh (1977) and Bourne (1977). Suppose a chimpanzee trained to use symbolic communication also engages in the preparation and use of probes to extract termites from termite nests, as observed by Goodall (1971, 1975). If an appropriate vocabulary had been taught to this chimpanzee, then while it prepared the probe (by stripping small twigs from a long thin stem of some appropriate plant), it could be asked, *What are you doing?* In response to such a question it might or might not indicate that it was contemplating the capture or eating of termites. In comparable fashion, other sorts of communicative behavior might or might not indicate the presence of internal representations of objects and events in the outside world, including those that the animal might be anticipating.

Chimpanzees that have learned sign language vocabularies of 100 words or more often appear to express desires or intentions; but relatively little investigation of this possibility has yet been reported. Washoe and her successors are routinely asked to name pictures or objects, or to choose between familiar activities for the immediate future. Savage-Rumbaugh et al. (1978) trained two chimpanzees to ask one another for specific types of food. It would be of great interest to ascertain the extent to which signing apes can communicate about objects and events that are remote in space or time. For instance, could they learn to use appropriately a sign meaning *tomorrow*? Could they describe plans for activities that cannot be carried out immediately? Could a chimpanzee state an intention to perform a sequential series of distinctly different acts, for example a plan that we might describe as follows: *If you open the door, I will go out, climb the big tree, pick an apple, and throw the apple at Roger?* The demonstration of such an ability would strongly indicate the presence within the animal's brain of a dynamic image of events and relationships remote from current sensory input, an internally formulated plan, and a structured series of intentions.

Testable hypotheses are also suggested by a cognitive approach to the communicative dances of honeybees. These are among the most clearly symbolic forms of naturally occurring animal communication known to ethologists (Frisch, 1967; Griffin, 1976). The waggle dances and accompanying sounds or near-field mechanical vibrations convey three types of information: distance, direction, and desirability. Chemical messages are also conveyed, including the scents of flowers or secretions from the forager's scent glands. The information conveyed by the waggle dances is probably accurate only to about ± 5° and ± 10% in distance, and odors enable recruits to locate the exact food source after reaching its general vicinity. The communication system is used for at least four commodities: sugar, pollen, water, and waxy materials. Furthermore, at the time of swarming, worker bees use it in a situation entirely new to their individual experience. No worker lives nearly as long as the normal interval between swarming episodes; but when thousands of bees have moved outside the cavity they have occupied for many worker life spans, some of the older workers that had been foraging for food search instead for cavities appropriate for establishing a new colony. Having located a cavity, such bees describe its distance, direction, and desirability with the same code (Lindauer, 1955, 1971). They mark the actual cavity with secretions from scent glands, which helps recruits find the exact spot. The same bee may act alternately as transmitter and receiver of information about cavities, and she may change from dancing about a mediocre cavity to describing a better one under the influence of another dancer whose message about the second cavity conveys the information that it is more suitable.

This exchange of messages and reaching of a consensus is such an unexpected form of behavior to find among insects that ethologists have almost totally ignored it in the 25 years since it was first described. It simply did not fit the Procrustean bed of our behavioristic *Zeitgeist,* any more than the experiments of Spallanzani and Jurine on auditory orientation of bats fitted into that of the 1790s (Griffin, 1958). Indeed, to the best of my knowledge, no one has even repeated Lindauer's original experiments, although Seeley (1977) recently studied the properties of cavities that lead to variations in the numbers of recruits.

One can interpret the application of the *Schwanzelntanz* to cavities as a precisely hard-wired, genetically programmed behavior pattern, never used for many generations, but ready, in latent form, to be elicited at the time of swarming. Suppose we allow ourselves to consider the alternative interpretations, that foraging bees know what they are doing. We might then describe the situation in the following terms, or in some roughly equivalent fashion: For several days bees have been communicating about nectar, which they know will be eagerly accepted by their hivemates. Now a cavity is obviously needed, scouts have searched and found one, and they are communicating its desirability and location with the intention that the colony move to this cavity. These exchanges of messages via waggle dances include many of the elements of a conversation, which seems to take the form of exchanging imperatives, but is one effective way to exchange ideas. And, as Lindauer described in detail, a consensus is reached after prolonged exchanges of such competing imperatives; only then does the entire swarm fly off to the cavity which has for some hours been described by the great majority of the dancers.

Some people are outraged at the suggestion that bees might *intend* to communicate to their sisters, but their basis for rejecting this possibility is not altogether clear. If comparable symbolic communication were observed in chimpanzees, we would infer intention with relatively little hesitation. Is it because bees are small, or because they are so distantly related to ourselves that we are so reluctant to allow them even the simplest of intentions? Are we allowing our thinking to be constrained by outmoded, pre-Frischian, concepts that all insect behavior is rigidly mechanical? This communicative behavior has many of the properties that we would expect if it were intentional. For instance, it occurs when and only when the dancer has discovered something needed by the colony, and many of the older workers have ample first-hand information about the needs of the colony, as described by Lindauer (1971). Smith (1977, p. 144) expressed a common view that honeybee dances communicate information "about characteristics of the next flight the dancing communicator will make" rather than about the location of something desirable; but the distinction between predicting one's future behavior and expressing an intention is a rather subtle one that is certainly difficult to analyze in another species. It is therefore appropriate to ask what else bees might be expected to do that would provide stronger evidence of intention to communicate, given the circumstances under which their behavior has been studied so far.

Relevant, although not totally conclusive evidence might be gathered from an experimental situation in which bees learn that something altogether new is needed by their colony. Under some circumstances they do gather and store unnatural materials such as tar; perhaps this happens because some component mimics the odors of natural foods, but too little is known to allow more than speculative interpretations of such behavior. Suppose bees could be trained to gather fiberglass particles, in order to produce some positive reinforcement from an experimental apparatus attached to an observation hive. Perhaps this could be a badly needed supply of sugar, poured directly into the hive, as is commonly done to maintain a weak colony unable to supply its own needs; or a desirable change of temperature might prove to be an appropriate reinforcement. If bees did learn that bringing back something as unnatural as fiberglass particles produced a major improvement in the conditions inside the hive, would they begin to dance about sources of this material? This would demonstrate communication about a relationship that the bees had learned de novo. While such experiments might be cumbersome, they are quite feasible in principle and would involve the objective testing of a well-defined hypothesis, namely, that worker bees can learn about new needs, supply them by gathering new sorts of material, and then communicate to their sisters the location and desirability of such materials.

As described by Lindauer, the alternation of transmitting and receiving roles of dancing bees reporting about cavities only involved cases in which the bee that changed her dances had first visited the cavity described by the more vigorous dances of a second bee. In other words, the sequence of events was approximately this: Bee N visits cavity n (north of the swarm) and dances vigorously about its location. At the same time bee W has been visiting a less desirable cavity w (located to the west) and has signified its distance and direction with correspondingly less vigorous dances. W then follows the dances of N and next flies out to cavity n. Only after doing so does she return to the swarm and dance about n with an intensity that apparently approximates that of bee N.

A behaviorist could explain what Lindauer reports in terms of competing influences on the worker bee: visiting a cavity with certain properties elicits a certain dance pattern; following a sufficiently vigorous dance elicits flying out in the indicated direction, to roughly the indicated distance, and searching for the odor accompanying the dance. If the second stimulus is sufficiently intense, it is more influential than the first and results in a third stimulus—visiting the superior cavity. This stimulus in turn elicits vigorous and appropriately oriented dances when the bee returns to the swarm. Our conventional *Zeitgeist* tends to dull our curiosity at this point and makes us satisfied to let the matter rest with this sort of interpretation.

The cognitive ethologist is likely, at the very least, to devise significant experiments of types that behaviorists are unlikely to contemplate. He might ask whether, after following the dances of N, bee W already has an internal replica of the message transmitted by N's dance, a message we might express approximately as *Very good cavity up north at about such-and-such a distance*. How could one hope to answer such a question? We know from the work of Frisch and his students that bees can remember the location of good food sources for many hours or even days. We know this because when bees are stimulated by the odor of a formerly rich food source that has been exhausted and abandoned they often fly out to it as though expecting that the flowers had again begun to open and make nectar or pollen available. Even in the middle of the night, when foraging never takes place, dances about a good food source can be elicited under special circumstances. Suppose, in the situation described above, we prevent bee W from leaving the hive, perhaps by timing the experiment so that darkness intervenes, or simply by blocking the entrance. Under any reasonable circumstances, would W dance about cavity n *before* she has visited it? If so, her communicative behavior would have been altered by following the dances of N, without the intervening step of visiting the cavity herself.

This straightforward experiment could certainly be carried out on a large enough scale to give a definite answer. Out of any reasonable number of replications, the bees designated W either would or would not dance about cavity n. Experience with animal behavior suggests the likelihood of a mixed result, namely, that some but not all Ws might dance about cavity n. However, even if only some individuals did so consistently, the experiment would show a direct effect of a received message on subsequently transmitted messages. Falsification, or at least negative evidence, would be provided by failure to dance about n without first visiting it.

The behavioristic ethologist could, of course, accept a positive result and interpret it as a shift in the pattern of stimulus-response relationship. However, behaviorists treat human thought and speech in the same fashion, and, therefore as pointed out above, their analysis is not discriminative in the important sense that it can be applied equally well to exchanges that we know involve intention and awareness and to those that almost certainly do not. For example, the same type of behavioristic analysis can be applied in a precisely parallel form to heat-seeking antiaircraft missiles, to bacteriophages, and to the most thoughtful human conversations.

A convincing demonstration of reported intentions would be all the more interesting the longer the time interval between the report of the intention and the performance of the behavior thus anticipated. In almost all known cases this interval seems to be only a few seconds or minutes, but perhaps this results from a lack of experimental interest in the possible occurrence of intentions concerning the more remote future. For example, do animals ever express anticipations of what they will do tomorrow or next week? Some behavior patterns develop gradually over days or weeks. These include preparations for migration or for mating; but our custom has been to interpret such cases as the gradual buildup of some physiological state leading to the full-fledged behavior in question. Do animals in these preliminary stages have any internal representations of the full-fledged behavior they will later carry out? When starting to construct nests or dig burrows, do they think in any way about the finished structure? Evidence for the presence of such mental images might be obtained from differential responses to pictures or other external representations of the finished product.

This brief list of hypotheses concerning reportable mental images and intentions could be expanded almost indefinitely. When an animal is suspected of experiencing a mental image or intention, cognitive ethologists may reasonably inquire whether it can report reliably about the postulated experience. What appears to be communication about an intention can be evaluated by observing how accurately it does, in fact, predict future behavior. In many cases the prediction may simply be an initial stage of the behavior in question. This is, of course, only weak evidence of a significant intention, since it may be nothing more than the beginning of a complex and prolonged response to the contemporary stimulus situation. For this reason, it is of special interest to investigate instances where animals may communicate about relatively remote objectives, that is, where their communication involves displacement in both space and time.

Signing apes and dancing bees have thus far dominated this discussion for the simple reason that their communications are richer and more strongly suggestive of underlying mental experiences than those yet described for other animals. Does this situation result from inadequate knowledge of other species, or are no others capable of equally complex communication? Cognitive ethologists may well be able to answer such questions, or at least narrow the range of uncertainty that now exists. Many groups of animals, such as songbirds and cetaceans, are known to emit signals that involve repeated patterns of varying complexity; but we do not know nearly enough about the messages actually conveyed. For example, the rich repertoire of humpback whale songs described by Payne and McVay (1971) and others is very suggestive; but the decoding of these signals has scarcely begun, and we do not know whether any of the various song patterns convey different messages. While the general opinion of ethologists is that they probably do not, it is appropriate to reflect how little anyone suspected, say, in 1920, what complex information is conveyed by the frenzied gyrations of dancing honeybees. A broad and general, but testable, hypothesis is that whenever the communication signals of a species include consistently distinguishable patterns, at least some of them convey different messages.

These suggestions are not intended to be a complete prescription of all the ways in which cognitive ethologists might hope to learn more about animal thinking. Other approaches will doubtless prove equally significant. Many complex learned discriminations and certain natural behavior patterns—such as constructing nests or burrows, dam building by beavers, and the like—would benefit from planning and may involve some degree of intention. A variety of other behavioral patterns may well be facilitated by understanding and insight on the animal's part; and students of animal behavior might profit both from inquiring as to what animals may be thinking when they engage in such behavior, and from thorough, imaginative, and critical analysis of the resulting hypotheses. Sociobiologists search for ways in which observed behavior may increase inclusive fitness, but direct rigorous testing of such hypotheses is so difficult that indirect evidence must be relied upon. Research in cognitive ethology will also require unraveling many tangled skeins, formulating hypotheses, and weighing all available evidence to evaluate them. However, the effort appears well justified when one contemplates the numerous surprises that have come from discoveries about animal behavior during the past 50 years.

Significance of Animal Awareness

To the extent that cognitive ethologists may gather convincing evidence that nonhuman animals do or do not have intentions, and are or are not aware of themselves in relation to their surroundings, our understanding of animals would be significantly enhanced. A given animal may or may not be aware of some object or relationship; if it is aware, we may or may not be able to obtain verifiable evidence of the fact. If we can do so, however, we will have established something very important about the animal in question. Furthermore, one clear and convincingly repeatable example demonstrates a capability, and other examples of the use of the capacity may be discovered once ethologists have learned where and how to look for them. Gallup's experiments discussed above are a good example of this type of progress.

If no species other than our own has any capability of self-awareness, intention, or other reportable types of mental experience, this is a most important fact to establish as conclusively as possible. Failure of cognitive ethologists to gather convincing evidence of mental experiences would, of course, constitute negative evidence, and negative evidence is never totally conclusive; but we are forced to rely on it to dismiss an infinite variety of unobserved phenomena as so unlikely that they warrant no consideration whatever. One can imagine six- or eight-legged mammals, but we do not waste time looking for them, although only negative evidence of their nonexistence is available. Animal awareness is in quite a different category because much animal behavior is consistent with (although it does not establish) the hypothesis that animals know what they are doing.

Ethical considerations concerning our treatment of animals are obviously affected by our views concerning animal awareness. Some followers of Descartes were so convinced that even dogs were unfeeling mechanisms that they vivisected them in ways from which even the strictest of 20th century behaviorists would recoil in horror. Postulating that some animals may be aware of some things and events need not convert one into a vegetarian. But improved understanding of animal awareness might lead to better informed decisions about the trade-offs between human benefits derived from animals and the effects of our exploitation on the animals concerned. Here we face a vast and difficult area of concern, an area of great philosophical significance, but one that lies far outside the scope of this chapter.

Animal awareness, if it occurs, is also important for our definition and understanding of the human condition. We have considered ourselves qualitatively unique, but evolutionary biology and ethology have revealed a continuity and kinship rather than all-or-nothing dichotomies. Cognitive ethology can illuminate the fundamental dimensions of those attributes we loosely call thinking and which, in their most versatile manifestations, are sources of our most profound satisfactions.

References

Bourne, G. H. (Ed.). *Progress in ape research*. New York: Academic Press, 1977.

Brown, J. W. Consciousness and the pathology of language. In R. W. Rieber (Ed.), *The neuropsychology of language, Essays in honor of Eric Lenneberg*. New York: Plenum Press, 1976.

Campbell, D. T., & Blake R. Animal awareness? *American Scientist*, 1977, *65*, 146-147.

Corballis, M. C., & Morgan, M. J. On the biological basis of human laterality: I. Evidence for a maturational left-right gradient. *Behavioral and Brain Sciences*, 1978, *1*, 261-269.

Dimond, S. J., & Blizard, D. A. (Eds.). Evolution and lateralization of the brain. *Annals of the New York Academy of Sciences*, 1977, *299*, 1-501.

Edwards, P., & Pap, A. (Eds.). *A modern introduction to philosophy* (3rd ed.). New York: Macmillan, 1973.

Fouts, R. S., & Rigby, R. L. Man-chimpanzee communication. In T. A. Sebeok (Ed.), *How animals communicate*. Bloomington, Ind.: Indiana University Press, 1977.

Frisch, K. von. *The dance language and orientation of bees.* (L. Chadwick, trans.) Cambridge, Mass.: Harvard University Press, 1967.

Galaburda, A. M., LeMay, M., Kemper, T. L., & Geschwind, N. Right-left asymmetries in the brain. *Science,* 1978, *199,* 852-856.

Gallup, G. G., Jr. Towards an operational definition of self-awareness. In R. Tuttle (Ed.), *Socio-ecology and psychology of primates.* The Hague: Mouton, 1975.

Gallup, G. G., Jr. Self-recognition in primates. A comparative approach to the bidirectional properties of consciousness. *American Psychologist,* 1977, *32,* 329-338.

Galusha, J. G., & Stout, J. F. Aggressive communication by *Larus glaucescens* Part IV: Experiments on visual communication. *Behaviour,* 1977, *62,* 222-235.

Gardner, B. T., & Gardner, R. A. Evidence for sentence constitutents in the early utterances of child and chimpanzee. *Journal of Experimental Psychology,* 1975, *104,* 244-267.

Globus, G. G., Maxwell, G., & Savodnick, I. *Consciousness and the brain.* New York: Plenum Press, 1976.

Goodall, J. van Lawick. *In the shadow of man.* Boston: Houghton Mifflin, 1971.

Goodall, J. van Lawick. The behaviour of the chimpanzee. In G. Kurth & I. Eibl-Eibesfeldt (Eds.), *Hominisation und Verhalten.* Stuttgart: Fischer, 1975.

Griffin, D. R. *Listening in the dark.* New Haven: Yale University Press, 1958. (Reprinted New York: Dover, 1974.)

Griffin, D. R. *The question of animal awareness: Evolutionary continuity of mental experience.* New York: Rockefeller University Press, 1976.

Griffin, D. R. Anthropomorphism. *BioScience,* 1977, *27,* 445-446. (a)

Griffin, D. R. Expanding horizons in animal communication behavior. In T. A. Sebeok (Ed.), *How animals communicate.* Bloomington, Ind.: Indiana University Press, 1977. (b)

Harnad, S. R., Steklis, H. D., & Lancaster, J. (Eds.). Origins and evolution of language and speech. *Annals of the New York Academy of Sciences,* 1976, *280.*

Haugeland, J. The nature and plausibility of cognitivism. *Behavioral and Brain Sciences,* 1978, *1,* 215-260.

Herrnstein, R. J., Loveland, D. H., & Cable, C. Natural concepts in pigeons. *Journal of Experimental Psychology: Animal Behavior Processes,* 1976, *2,* 285-302.

Hölldobler, B. Communication in social hymenoptera. In T. A. Sebeok (Ed.), *How animals communicate.* Bloomington, Ind.: Indiana University Press, 1977.

Humphrey, N. K. Review of *The question of animal awareness. Animal Behaviour,* 1977, *25,* 521-522.

Jenssen, T. A. Female response to filmed displays of *Anoblis nebulosus* (Sauria, Iguanidae). *Animal Behaviour,* 1970, *18,* 640-647.

Kenny, A. J. P., Longuet-Higgins, H. C., Lucas, J. R., & Waddington, C. H. *The development of mind.* Edinburgh: Edinburgh University Press, 1973.

Kenny, A. J. P., Longuet-Higgins, H. C., Lucas, J. R., & Waddington, C. H. *The nature of mind.* Edinburgh: Edinburgh University Press, 1972.

Krebs, J. Review of *The question of animal awareness. Nature,* 1977, *266,* 792.

Kupfermann, I., & Weiss, K. R. The command neuron concept. *Behavioral and Brain Sciences,* 1978, *1,* 3-39.

Lashley, K. S. The behavioristic interpretation of consciousness. *Psychological Review,* 1923, *30,* 237-272, 329-353.

Lindauer, M. Schwarmbienen auf Wohnungssuch. *Zeitschrift für vergleichenden Physiologie,* 1955, *35,* 263-324.

Lindauer, M. *Communication among social bees* (Rev. Ed.). Cambridge, Mass.: Harvard University Press, 1971.

Mackenzie, B. D. *Behaviourism and the limits of scientific method.* London: Routledge and Kegan Paul, 1977.

Marler, P., & Tenaza, R. Signaling behavior in apes with special reference to vocalization. In T. A. Sebeok (Ed.), *How animals communicate.* Bloomington, Ind.: Indiana University Press, 1977.

Mason, W. A. Review of *The question of animal awareness. Science,* 1976, *194,* 930-931.

Medin, D. L., Roberts, W. A., & Davis, R. T. (Eds.). *Processes of animal memory.* New York: Wiley, 1976.

Menzel, E. W., & Halperin, S. Purposive behavior as a basis for objective communication between chimpanzees. *Science,* 1975, *189,* 652-654.

Menzel, E. W., & Johnson, M. K. Communication and cognitive organization in humans and other animals. *Annals of the New York Academy of Sciences,* 1976, *280,* 131-142.

Neville, H. J. The functional significance of cerebral specialization. In R. W. Reiber (Ed.), *The neuropsychology of language, Essays in honor of Eric Lenneberg.* New York: Plenum Press, 1976.

Nicholas, J. M. (Ed.). *Images, perception, and knowledge.* Dordrecht, Holland: Reidel, 1977.

Nottebohm, F. Neural asymmetries in the vocal control of the canary. In S. R. Harnad & R. W. Doty (Eds.), *Lateralization in the nervous system.* New York: Academic Press, 1977.

Payne, R. S., & McVay, S. Songs of the humpback whales. *Science,* 1971, *173,* 587-597.

Popper, K. R., & Eccles, J. C. *The self and its brain.* New York: Springer-Verlag, 1977.

Premack, D. *Intelligence in ape and man.* Hillsdale, N.J.: Erlbaum, 1976.

Puccetti, R., & Dykes, R. W. Sensory cortex and the mind-brain problem. *Behavioral and Brain Sciences,* 1978, *1,* 337-375.

Riopelle, A. J. (Ed.). *Animal problem solving.* Hammondsworth, England: Penguin Books, 1967.

Roland, P. E. Sensory feedback to the cerebral cortex during voluntary movement in man. *Behavioral and Brain Sciences,* 1978, *1,* 129-171.

Rumbaugh, D. M. (Ed.). *Language learning by a chimpanzee.* New York: Academic Press, 1977.

Ryle, G. *The concept of mind.* London: Hutchinson, 1949.

Savage-Rumbaugh, E. S., Wilkerson, B. J., & Bakeman, R. Spontaneous gestural communication among conspecifics in the pigmy chimpanzee (*Pan paniscus*). In G. H. Bourne (Ed.), *Progress in ape research.* New York: Academic Press, 1977.

Savage-Rumbaugh, E. S., Rumbaugh, D. M., & Boysen, S. Symbolic communication between two chimpanzees (*Pan troglodytes*). *Science,* 1978, *201*, 641-644.

Schaffer, J. A. Philosophy of mind. *Encyclopedia Brittanica Macropedia* (Vol. 12). Chicago: Encyclopedia Brittanica, 1975, pp. 224-233.

Schultz, D. *A history of modern psychology* (2nd ed.). New York: Academic Press, 1975.

Searle, J. R. *Speech acts, an essay in the philosophy of language.* London: Cambridge University Press, 1969.

Sebeok, T. A. (Ed.). *How animals communicate.* Bloomington, Ind.: Indiana University Press, 1977.

Seeley, T. Measurement of nest cavity volume by the honey bee (*Apis mellifera*). *Behavioral Ecology and Sociobiology,* 1977, *2*, 201-227.

Segal, S. J. (Ed.). *Imagery, Current cognitive approaches.* New York: Academic Press, 1971.

Sheehan, P. W. (Ed.). *The function and nature of imagery.* New York: Academic Press, 1972.

Shepard, R. N. The mental image. *American Psychologist,* 1978, *33*, 125-137.

Skinner, B. F. *Verbal behavior.* New York: Appleton-Century-Crofts, 1957.

Skinner, B. F. *About behaviorism.* New York: Random House, 1974.

Smith, W. J. *The behavior of communicating.* Cambridge, Mass.: Harvard University Press, 1977.

Taylor, J. G. *The behavioral basis of perception.* New Haven: Yale University Press, 1962.

Thatcher, R. W., & John, E. R. *Foundations of cognitive processes.* Hillsdale, N.J.: Erlbaum, 1977.

Thorpe, W. H. *Learning and instinct in animals* (2nd ed.). Cambridge, Mass.: Harvard University Press, 1963.

Whiteley, C. H. *Mind in action, an essay in philosophical psychology.* London: Oxford University Press, 1973.

Wilson, E. O. *The insect societies.* Cambridge, Mass.: Harvard University Press, 1971.

Author Index

Subject Index